Perinatal Brain Development, Malformation, and Injury

Colloquium Series on The Developing Brain

Editors

Margaret McCarthy, Ph.D.

Professor and Associate Dean for Graduate Studies
Department of Physiology
University of Maryland School of Medicine
Baltimore, Maryland

The goal of this series is to provide a comprehensive state-of-the art overview of how the brain develops and those processes that affect it. Topics range from the fundamentals of axonal guidance and synaptogenesis prenatally to the influence of hormones, sex, stress, maternal care and injury during the early postnatal period to an additional critical period at puberty. Easily accessible expert reviews combine analyses of detailed cellular mechanisms with interpretations of significance and broader impact of the topic area on the field of neuroscience and the understanding of brain and behavior.

My research program focuses on the influence of steroid hormones on the developing brain. During perinatal life, there is a sensitive period for hormone exposure during which permanent cytoarchitechtural changes are established. Males and females are exposed to different hormonal milieus and this results in sex differences in the brain. These differences include alterations in the volumes of particular brain nuclei and patterns of synaptic connectivity. The mechanisms by which sexually dimorphic structures are formed in the brain remains poorly understood.

I received my PhD in Behavioral and Neural Sciences from the Institute of Animal Behavior at Rutgers University in Newark, NJ in 1989. I then spent three years as a post-doctoral fellow at the Rockefeller University in New York, NY and one year as a National Research Council Fellow at the National Institutes of Health, before joining the faculty at the University of Maryland. I am a member of the University of Maryland Graduate School and the Center for Studies in Reproduction I am also a member of the Society for Behavioral Neuroendocrinology, the Society for Neuroscience, the American Physiological Association and the Endocrine Society.

Perinatal Brain Development, Malformation, and Injury
Juliet K. Knowles and Anna A. Penn
www.morganclaypool.com

ISBN: 9781615043422 paperback

ISBN: 9781615043439 ebook

DOI: 10.4199/C00044ED1V01Y201109DBR006

A Publication in the

COLLOQUIUM SERIES ON THE DEVELOPING BRAIN

Lecture #6

Series Editor: Margaret M. McCarthy, University of Maryland School of Medicine

Series ISSN

ISSN 2159-5194   print
ISSN 2159-5208   electronic

Perinatal Brain Development, Malformation, and Injury

Juliet K. Knowles
Lucile Packard Children's Hospital
Department of Pediatrics
Stanford University School of Medicine

Anna A. Penn
Division of Neonatal and Developmental Medicine
Department of Pediatrics
Stanford University School of Medicine

COLLOQUIUM SERIES ON THE DEVELOPING BRAIN #6

MORGAN & CLAYPOOL LIFE SCIENCES

ABSTRACT

We provide a broad overview of human brain development with associated malformations and injuries that occur in the period between early embryogenesis and delivery. The aim of this review is to summarize current understanding of the molecular and environmental cues that shape the developing brain. For each developmental stage, we give examples of disorders that arise from genetic and/or environmental insults to illustrate critical points of neurological susceptibility.

KEYWORDS

fetal brain, neurodevelopment, neural tube, neurogenesis, gliogenesis, placenta, developing cortex

Contents

1. Introduction 1

2. Origin of the Central Nervous System: The Neural Tube (3–4 Weeks Gestation).... 3
   2.1 Neurulation 3
       2.1.1 Molecular Mechanisms of Neurulation 5
       2.1.2 Molecular Mechanisms of Neural Tube Closure: Wnts and
             Hedghog Proteins 7
       2.1.3 Neural Tube Defects 10
       2.1.4 Etiology of Neural Tube Disorders: Genetics and
             Environment Collide 12
             2.1.4.1 Shh and Wnt Mutations 12
       2.1.5 Folate Metabolism 13
   2.2 Summary: Neurulation and NTDs 15

3. Patterning of the Neural Tube: A Blueprint for the CNS 17
   3.1 Dorso-Ventral Patterning of the Neural Tube 17
   3.2 Rostro-Caudal Patterning of the Neural Tube... 17
   3.3 Prosencephalic Cleavage and Midline Development................................. 20
   3.4 Midline Defects: Holoprosencephaly, Agenesis of the Corpus Callosum, and
       Dandy–Walker Malformation 21
       3.4.1 Holoprosencephaly 21
       3.4.2 Agenesis of the Corpus Callosum 23
       3.4.3 Dandy–Walker Malformation 28
   3.5 Genetics and Environment Collide Again to Generate These Midline
       Alterations 29
   3.6 Summary: CNS Patterning and Midline Defects..................................... 30

4. Neural Proliferation and Migration (3 Months Gestation Into Postnatal Period).. 31
   4.1 Neurogenesis 31
   4.2 Programmed Cell Death 33

4.3   Microcephaly vera: Deficient Neurogenesis..34
4.4   Neuronal Migration..35
4.5   Disordered Migration: Lissencephaly and LIS1..37
4.6   Summary: Neural Proliferation and Migration 38

5.   **Organization of Neuronal Circuits and Synaptogenesis (5 Months Gestation–Postnatal Years)** **41**
    5.1   Wiring Up the Brain: Axon Path Finding Between Brain Regions 41
    5.2   The Role of Subplate Neurons in Defining the Gross Structure of Circuits..... 43
    5.3   Synaptogenesis and Critical Periods: Fine Control of Circuits 45
    5.4   Immune Molecules and Circuit Refinement 47
    5.5   Impaired Synapse Formation: Fragile X 48
    5.6   Summary: Neuronal Circuits and Synaptogenesis.. 50

6.   **Gliogenesis and Myelination (5-Month Childhood)** **51**
    6.1   Glial Proliferation..51
    6.2   Astrogenesis...51
    6.3   Oligogliogenesis and Myelin Formation 51
    6.4   Microglia..53
    6.5   Summary: Gliogenesis..53

7.   **Developmental Brain Injury: Before, During, and After Birth** **55**
    7.1   Neonatal Encephalopathy..55
    7.2   Risk Factors for Injury Before Birth: Compromised Placental Function and Infection 57
    7.3   Injury in the Preterm Brain 58
        7.3.1   Periventricular Leukomalacia 58
        7.3.2   Intraventricular Hemorrhage.. 60
    7.4   Injury in the Term Brain...61
        7.4.1   Hypoxia–Ischemia 61
        7.4.2   Stroke 62
    7.5   Summary: Brain Injury..63

8.   **Conclusion** **65**

References **67**

Author Biographies...**73**

CHAPTER 1

Introduction

In the current era of specialized high-risk obstetrics and neonatal intensive care units, many new-borns born extremely prematurely or term newborns with complex brain injuries survive the neonatal period. However, morbidity from early brain injury is substantial, with enormous impact on the lives of children and their families. This monograph provides a broad overview of perinatal brain development and injury, from conception to birth, with a focus on disorders that may arise from genetic and/or environmental insults at each developmental stage.

The development of the human brain may be broadly divided into five stages (Volpe, 2008).

1. **0–1 month gestation:** formation of a neural plate (neurulation) and then the neural tube, which will give rise to the brain and spinal cord.
2. **2–3 months gestation:** development of the prosencephalon, which will form the basic structure of the brain and the cerebellum.
3. **3–5 months gestation:** neurogenesis and neuronal migration.
4. **5 months gestation to adulthood:** circuit organization.
5. **Birth to postnatal years:** myelination.

By concisely describing each developmental stage along with corresponding disorders, we aim to provide a clinical context for current research and to inspire students embarking on a career in neuroscience to direct their energies to this important field.

· · · ·

CHAPTER 2

Origin of the Central Nervous System: The Neural Tube (3–4 Weeks Gestation)

2.1 NEURULATION

Before the nervous system is formed, all vertebrate embryos begin as bi-layered ball of cells (the blastocoel) which invaginates (a process referred to as gastrulation) to form three germ cell layers: ectoderm (the outer layer), mesoderm (the middle layer) and endoderm (the inner layer). Each layer gives rise to distinct organs.

As shown in Figure 1, the tissue that will form the central nervous system (CNS) arises from ectodermal tissue on the dorsal aspect of the embryo (Ten Donkelaar and Hori, 2006). The notochord and prechordal plate sit under the ectoderm and send signals that transform an area of ectoderm into the neural plate, a specialized disc of neural tissue. Within the neural plate, the floorplate (on the "floor" of the neural tube) is induced just dorsal to the notochord (Purves et al., 2001; Ten Donkelaar and Hori, 2006). Once formed, the neural plate must roll up to form the neural tube. The lateral margins of the plate begin to curl and subsequently fuse to form the neural tube. Fusion begins at a mid-point in the area of the brainstem and simultaneously proceeds rostrally (toward the anterior pole of the embryo, from which the head will form) and caudally (toward the inferior pole where the spinal cord will end); finally, both ends of the tube (the anterior and posterior neuropores) fuse to form a closed structure (Volpe, 2008). Closure of the anterior neuropore occurs by embryonic day 24, while closure of the posterior pore occurs by approximately day 26 (equivalent to approximately 5 1/2 weeks gestation when counted using medical conventions). Neural tube disorders (NTDs) represent complete or partial failure of the neural tube to form and close (see below).

The formation of a neural tube from undifferentiated ectoderm takes approximately 1 week. The rostral end will subsequently give rise to the brain, while the caudal end of the tube forms most of the spinal cord. The terminal portion of the spinal cord, located distal to the upper sacral levels, including the conus and filum terminale, is formed from an undifferentiated cell mass at the caudal tip of the neural tube in a separate and later process known as secondary neurulation. Secondary neurulation involves tube formation via cavitation and elongation of a solid group of cells rather than a curling process. This second process begins at 4–7 weeks gestation; differentiation of these structures may not be complete until after birth (Volpe, 2008).

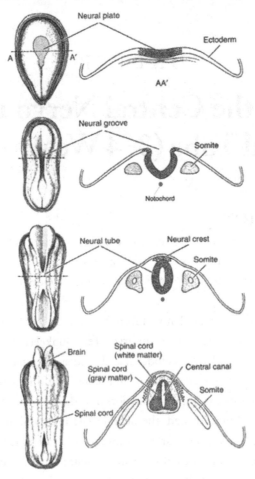

FIGURE 1: Schematic of neurulation. Formation of the neural plate, neural tube and neural crest. External view (left), Cross section (right). Floorplate is purple, roofplate is green, notochord is red. See text for details. Modified with permission from Elsevier (Volpe 2008).

As the two edges of the neural tube fuse, neural crest cells are formed (Figure 1). These cells subsequently migrate away from the developing neural tube and give rise to distinct, highly diverse components of the CNS and PNS, including the dorsal root ganglia, sensory ganglia of the cranial nerves, autonomic ganglia, Schwann cells, as well as the pia and arachnoid mater tissues that cover the CNS. Other neural crest cells become the neurosecretory cells of the adrenal glands as well as non-neural tissues, such as melanocytes. Soon after the neural tube closure, a group of cells in the dorsal area of the tube are induced to become the roof plate (Figure 1, green) (Purves et al., 2001).

At each step in this process, a complex interplay of molecular mechanisms leads to the ultimate formation of the brain and spinal cord. The keys to understanding how neurulation happens—normally or abnormally—lie in the molecular underpinnings of this complex process.

2.1.1   Molecular Mechanisms of Neurulation

How does the neural plate, a circumscribed area of initially undifferentiated ectoderm, become a neural tissue? In the developing embryo, the developmental fate that a particular cell will adopt

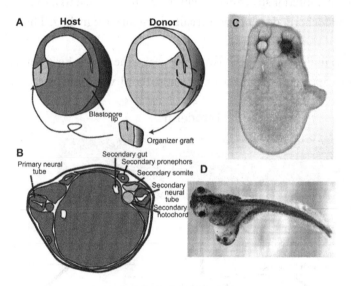

Harland R Development 2008;135:3321-3323

FIGURE 2: Spemann-Mangold Organizer. Experiments in which the organizer tissue from one amphibian embryo is transplanted to another indicated that there must an induction factor that could lead to the formation of a second neural axis. (A) light-gray newt donor organizer region grafted into a dark-gray host. (B) The famous result of an optimal grafting experiment (Spemann and Mangold, 1924), showing a section through the trunk of a twinned embryo. The light-gray graft has contributed to the notochord, medial somite and floor plate of the secondary axis. The graft has an induced neural tube, somites, a pronephros and a secondary archenteron cavity. (C, D) Contemporary organizer grafts from Andrea E. Wills (UC Berkeley, CA, USA). (C) The section shows a rafted organizer labeled with *lacZ* mRNA and stained with Red-Gal; the section is taken through the trunk of a stage 28 *Xenopus laevis* embryo, where the axial tissues are also stained with Tor70 antibody. (D) Twinned *Xenopus* embryo, resulting from an organizer graft carried out at stage 10. Reproduced with permission from Company of Biologists (UK) (Harland 2008).

is highly dependent on the presence of molecular inducing factors, secreted by cells elsewhere in the embryo, and the ability of a particular cell to respond to these inducing factors (Kandel et al., 2000).

In the case of the nervous system, the original inducing factors required for neurulation derive from a specialized group of cells known as the Spemann–Mangold organizer (named for the scientists who discovered it in 1924). Under the direction of molecular signals from the organizer region, the head, trunk and tail are formed (Niehrs, 2004). As depicted in Figure 2, transplantation of the organizer tissue from one amphibian embryo can lead to the formation of a second, conjoined "twin" embryo at the site of the transplant (Spemann and Mangold, 2001; De Robertis and Kuroda, 2004). Along with other interior structures, a cylinder of mesodermal tissue running the axial length of the embryo, the notochord, develops as a result of organizer signaling. This structure will go on to induce formation of the neural plate.

How does this induction happen? For many years, the search for a neural inducer molecule emanating from the Spemann organizer or the notochord was fruitless. A recent breakthrough

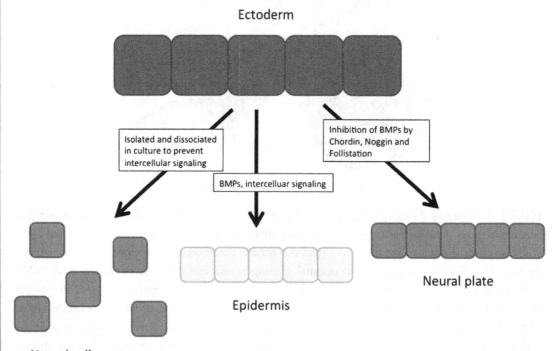

FIGURE 3: Effects of BMPs and Chordin, Noggin and Follistatin on ectoderm. The experiments shown schematically here demonstrated the striking role of *inhibitors* of BMPs (Chordin, Noggin and Follistatin) in allowing neural plate to form. Ectoderm exposed to BMPs becomes epidermis (skin). Original drawing by author.

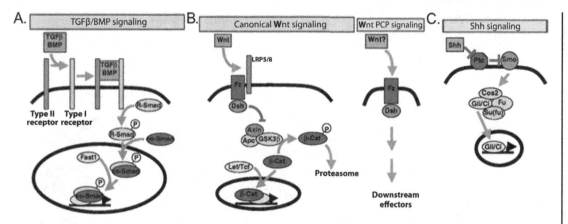

FIGURE 4: Schematic of key signaling molecule pathways. (A) Bone morphogenic proteins (BMP) (B) Wnts (C) Sonic hedgehog (Shh) See text for details. Modified with permission from Company of Biologists (UK) (Charron and Tessier-Lavigne 2005).

came from culturing undifferentiated embryonic ectoderm as single cells. In this "default" culture, the isolated cells differentiated into neural tissue (De Robertis and Kuroda, 2004) (Figure 3, left). This finding suggested that inducing factors were necessary for the differentiation of ectoderm in epidermal (skin) tissues, not neural tissue. These inducing factors, as identified in *Xenopus laevis* (frog) embryos, are a subclass of transforming growth factor-β (TGFβ) signaling molecules known as bone morphogenic proteins (BMPs) (Figure 4A). Exposing ectoderm to BMPs makes it differentiate into skin (De Robertis and Kuroda, 2004) (Figure 3, middle). But why doesn't the region destined to become neural tissue see these epidermis-inducing factors? It turns out that the organizer region and the notochord also secrete BMP inhibitors. These inhibitors, including Noggin, Chordin and Follistatin (Kandel et al., 2000), are secreted molecules made at the appropriate place and time to block the action of BMPs, allowing the formation of neural tissue from ectoderm (Figure 3, right) (Ten Donkelaar and Hori, 2006).

2.1.2   Molecular Mechanisms of Neural Tube Closure: Wnts and Hedgehog Proteins

After neural plate induction, what makes the plate fold into a tube? First, the neural plate morphs from an ovoid to a keyhole shape; the narrow part is destined to become the spinal cord while the enlarged end will become the brain (Copp et al., 2003) (Figure 5). This change in shape requires yet another class of signaling molecules known as Wnts (Figure 4B). Wnt molecules are secreted glycoproteins which play essential roles in virtually every stage of brain development. These molecules can signal in multiple ways, but for medial movement of cells in the caudal (narrower) portion of

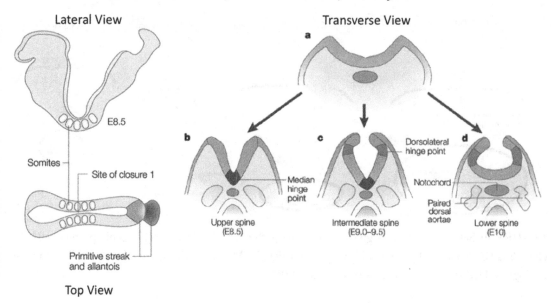

FIGURE 5: Embryonic Neural Tube Folding, Bending, Closing (Mouse). Lateral and top view of mouse embryo at embryonic day (E) 8.5 showing a well-defined anterior neural plate that is flanked by five pairs of somites. The neural folds at the level of the third somite pair approach each other in the midline, to create the incipient closure 1 site (creating the anterior and posterior neuropores that will close as the tube "zips" from the middle). The transverse view shows (a) neural fold bending (b) neural fold elevation varies in morphology along the body axis, with bending solely at the median hinge point (MHP) at upper levels of the body axis (c) at both the MHP and the paired DORSOLATERAL HINGE POINTS (DLHPs) at intermediate spinal levels and (d) DLHPs at the lowest spinal levels. Modified with permission from Nature Publishing Group (Copp et al., 2003).

the neural plate that will become the spinal cord ("convergent extension"), the Wnt pathway seems to work by inducing cytoskeletal remodeling via Rho GTPases (using a pathway known as the non-canonical pathway in contrast to the "canonical" signaling via Frizzled, a G-protein coupled receptor, Figure 4b) (Freese et al., 2010 ; Cohen, 2003; Copp et al., 2003; Salie et al., 2005; Lindwall et al., 2007).

The lateral edges of the plate curl until they meet at the center and fuse. The closed portion of tube is then covered by the ectodermal tissue which is used to flank the neural plate (Copp et al., 2003). Curling requires the emergence of a medial hinge point and paired dorsolateral hinge points along the rostrocaudal axis of the neural plate (Figure 5). The development of these "kinks" in the neural tube requires the proper signaling of another conserved and oft-recurring pathway initiated by Sonic hedgehog (Shh) protein (Figure 4c). Shh is secreted by the notochord, and later by the floor plate. It diffuses to the ventral neural tube to and binds to a receptor, Patched, within the target cells of the neural tube, resulting in the release of a protein, Smoothened, and in the increased activity of transcription factors in the Gli family (Traiffort et al., 2010; Salie et al., 2005) (Figure 4C).

A specialized cellular structure, the primary cilium, has recently been recognized as a critical structure for many of these signaling pathways. Primary cilia are highly evolutionarily conserved microtubule-based cellular organelles found on the apical surface of most cells during embryonic development. A key feature of primary cilia is their ability to conduct intraflagellar transport. Primary cilia are now thought to act as chemical and mechanical receptors that cause the cell to respond to developmental signals appropriately, particularly the Shh pathway, and possibly the Wnt pathway (Huangfu et al., 2003; Huangfu and Anderson, 2005) (Figure 6). Shh receptor, Patched, has been found on the cilium. When intraflagellar transport machinery is blocked or normal primary cilia are eliminated in mice, the mice have similar deficits to those seen in mice where Shh signaling has been genetically enhanced by removal of patched, its negative regulator (Huangfu et al., 2003).

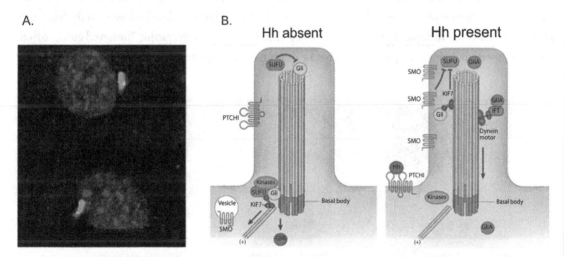

FIGURE 6: Primary cilia as site of signaling. A. Embryonic mammalian cells have primary cilia (green). The basal bodies (red) sit at the base of the cilia, nuclei are stained blue. B. Schematic of Shh signaling in mammalian cilia. In the absence of Shh, PTCH1 localizes to the cilium , blocking the entry of SMO into cilia. The kinesin KIF7 localizes to the base of the cilium, where it may form a complex with Gli proteins and other pathway components. KIF7 at the cilium base prevents Gli enrichment within the cilium and promotes processing of GliRs. In the presence of Shh, SMO moves to the ciliary membrane and KIF7 translocates into the cilium, thereby promoting GLI2 accumulation at the cilium tip. Activated Gli is then transported out of the cilium by the dynein motor and intraflagellar transport (IFT) particles. Modified with permission from Nature Publishing Group (Goetz and Anderson 2010).

2.1.3 Neural Tube Defects

When the processes of neurulation and neural tube closure are disrupted, it results in a group of disorders called *neural tube defects* (NTDs). NTDs are the most prevalent type of congenital malformation after heart defects with a worldwide incidence of 1:1000 pregnancies, (Copp et al., 2003). There is not only a strong genetic component, reflected in familial recurrence risk of 2–5% for NTDs in subsequent pregnancies, but also a large environmental component, with dietary factors, toxic exposures and hyperthermia all increasing risk as well.

NTDs vary in severity depending on the point at which the disruption occurs (Figure 7). For example, failure to induce a neural plate results in craniorachischisis totalis (complete absence of the neural tube and thus the brain and spinal cord, Figure 7A). More commonly, failure of the anterior portion of the tube to close results in anencephaly (Figure 7B), complete absence of the brain, while partial failure of anterior closure may result in encephalocele (Figure 7C), in which part of the brain herniates through a skull defect. At the posterior end, failure of the of the tube to close leads to spina bifida (Figure 7C), the most common NTD, in which complete or partial closure failure leads to spinal cord abnormalities (myeloschisis or myelomeningocele, respectively) (Fenichel, 2007). In the subtlest form, failure of the caudal portion of the spinal cord (which develops separately from the more superior cord) to separate from surrounding tissue results in a dysraphic "tethered cord," often marked superficially by a deep sacral dimple or dermal sinus (Figure 8).

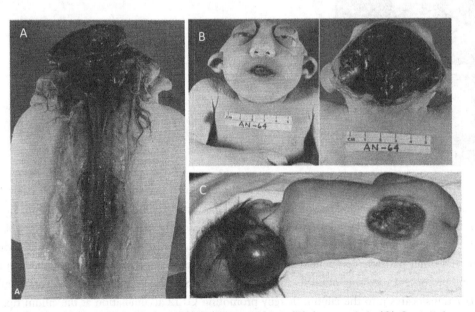

FIGURE 7: Severe Neural Tube Defects. (A) Craniorachischsis (B) Anencephaly (C) Occipital encephalocele and a thoracolumbar myelomeningocele. Reproduced with permission from Elsevier (Volpe 2008).

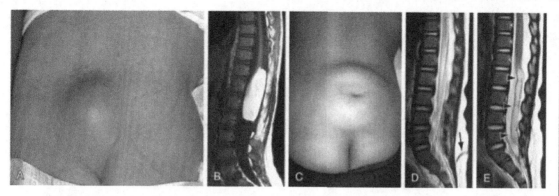

FIGURE 8: Occult Neural Tube Defects. Clinical features and imaging findings associated with occult spinal dysraphism. A, Spinal lipoma in a 2 mo old girl. There is a skin-covered lumbosacral mass above the intergluteal cleft in the midline, surrounded by overgrowing hair. B, Sagittal T1-weighted image shows huge intradural lipoma, merging with the conus medullaris superiorly. C, Lipoma and central dermal sinus. D and E, Dermal sinus with dermoid, 8 yr old girl. Slightly parasagittal T2-weighted image shows sacral dermal sinus coursing obliquely downward in subcutaneous fat (*arrow*) (D). Midsagittal T2-weighted image shows huge dermoid in the thecal sac (*arrowheads*), extending upward to the tip of the conus medullaris (E). The mass gives a slightly lower signal than cerebrospinal fluid and is outlined by a thin low-signal rim. Reproduced with permission from Elsevier (Kliegman 2011).

NTDs are commonly associated with other brain malformations. For instance, myelomeningocele is usually accompanied by the constellation of abnormalities which are referred to as a Chiari II malformation (downward displacement of the medulla, fourth ventricle and lower cerebellum into the cervical canal, abnormal elongation of the pons and medulla, and abnormalities of the surrounding bony structures shown in Figure 9. This malformation is particularly problematic as it can lead to dysfunction of the parts of the brain stem which control breathing and feeding (Volpe, 2008).

Severe NTDs such as anencephaly are lethal; in milder forms such as myelomeningocele, motor functions are usually impaired. Posterior NTDs such as myelomeningocele are most common and affected individuals have motor and sensory deficits in the legs, urinary and fecal incontinence, and hydrocephalus (buildup of cerebrospinal fluid which exerts pressure on the brain) (Copp et al., 2003). Surgical management includes closure of the defect at birth and later ventricular shunting to relieve hydrocephalus which can improve neurologic function. Fetal surgery has recently been shown to reduce exposure of the cord to amniotic fluid and decrease the rate of postnatal hydrocephalus, but remains experimental given the significant risks to both the fetus and mother (Adzick et al., 2011; Figure 9).

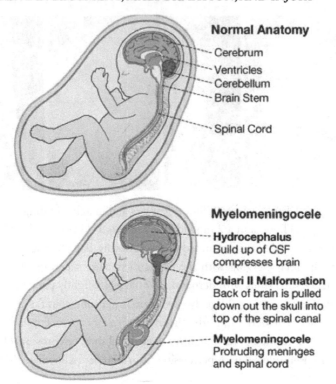

FIGURE 9: Conditions associated with neural tube defects. Hydrocephalus and Chiari II Malformation. Reproduced with permission from UCSF. http://fetus.ucsfmedicalcenter.org/spina_bifida/

2.1.4 Etiology of Neural Tube Disorders: Genetics and Environment Collide

Modes of inheritance of NTDs may include genetic and/or environmental causes. In the majority of cases, NTDs, and indeed almost all CNS malformations, are complex traits, meaning that both environmental and genetic causes are at work. NTDs are most often non-syndromic, occur with incomplete penetrance, and likely involve networks of multiple genes (Blom et al., 2006).

2.1.4.1 Shh and Wnt Mutations. Given the roles that Shh and Wnt signaling play in neurulation and neural tube closure, it is not surprising that mutations in these pathways cause NTDs. As described previously, Wnt signaling is required for convergent extension that precedes tube formation. Mice with targeted mutations affecting various molecular steps in the non-canonical Wnt cascade exhibit craniorachischisis totalis—total failure of neural tube formation. For example, the *crash* mouse carries a mutation in *Celsr1*, the orthologue of the Drosophila gene *flamingo*, en-

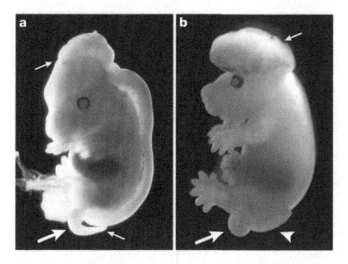

FIGURE 10: Mouse mutant mimics human neural tube defects. (a) craniorachischisis in a *Celsr1* mutant and (b) exencephaly and open spina bifida in a *curly tail* (ct) mutant. See text for details. Reproduced with permission from Nature Publishing Group (Copp, Greene et al. 2003).

coding a protein that interacts with the Wnt receptor, Frizzled (Figure 10a). Other mutations create more limited defects; for example those that increase Shh signaling lead to more localized neural tube defects that resemble human myelomeningocele (Figure 10b) (Copp et al., 2003).

Mutations which prevent normal primary cilium function have been shown to result in over-activity of Shh, again resulting in NTDs. In a human example, Meckel–Gruber syndrome is a rare, single gene disorder that disrupts cilia and is characterized by: posterior encephalocoele (a particular type of NTD); cleft palate and lip; cystic malformations of multiple internal organs; and polydactyly. The role of ciliopathies in CNS malformations is only now becoming appreciated, leaving many opportunities for new discoveries and possible therapies related to NTDs (Logan et al., 2011).

2.1.5   Folate Metabolism

While genetics clearly plays a role in NTDs, most cases or sporadic and non-syndromic (>95%) and appear to arise from an interplay between genetic predispositions and the environment of the developing embryo. From a public health success standpoint, the most recognizable dietary factor is the decrease in NTDs when folate (a B vitamin) intake is increased through supplementation in foods or through prenatal vitamins. Folate normally functions as a single carbon unit (methyl group)

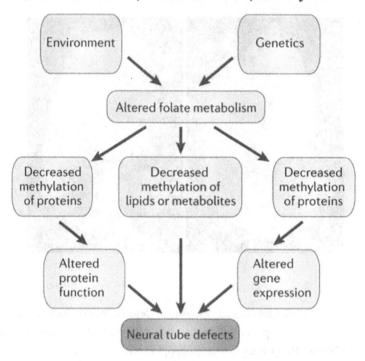

FIGURE 11: Interplay of environment and genetics associated with folate sensitive neural tube defects. Reproduced with permission from Nature Publishing Group (Blom, Shaw et al. 2006).

donor in methylation reactions required for biochemical reactions involving proteins, lipids and other metabolites. Some maternal genetic mutations linked with fetal NTDs have decreased methyl donor activity (Blom, Shaw et al. 2006). This has led to the hypothesis that impairment of methylation reactions may underlie NTDs (Figure 11). In 1991, a randomized controlled clinical trial demonstrated that folate supplementation prevented 70% of NTDs, leading to the universal recommendation that women who are planning to become pregnant take folic acid supplements and, in 1998, to fortification of the U.S. food supply with folic acid (Blom, Shaw et al. 2006; Greene et al., 2009).

There are rare causes of NTDs linked to folate deficiency—consumption of drugs that act as folate antagonists such as anti-epileptics valproate and carbemazepine, maternal autoimmune production of antibodies directed against folate receptors, and genetic abnormalities in folate metabolism—but the majority have no clear link to folate status, despite the preventive effect of supplementation. Furthermore, folate supplementation can prevent NTDs only in some of the mouse models of NTDs. Thus, many questions remain about how folate prevents NTDs and why supplemental folate can prevent NTDs in some cases, but not others.

2.2   SUMMARY: NEURULATION AND NTDs

- Neurulation is the embryonic stage in which primitive ectoderm forms into the neural tube, a structure that will give rise to the CNS.
- Multiple, precisely coordinated positive and negative molecular signals are required for neural tube formation.
- Failures in tube closure leads to NTDs, most commonly anencephaly (anterior failure) and spina bifida (posterior failure).
- Causes of NTDs are multifactorial (genetic, metabolic, environmental).
- Taking folate prior to conception and during pregnancy is the major intervention shown to reduce the risk of NTDs.

·   ·   ·   ·

CHAPTER 3

Patterning of the Neural Tube: A Blueprint for the CNS

As the neural tube forms, its cellular components begin to differentiate into specialized tissues. This regionalization, or patterning, occurs along two axes: dorso-ventral and rostro-caudal (Figure 12) Gradients of signals along these two axes cause neural tube cells to take on different characteristics by triggering a variety of gene expression patterns in groups of cells exposed to specific combinations of secreted molecules. Ultimately, the human brain will be made up of tens of billions of neurons that can be roughly divided into more than 5000 different, regionalized cell types; this basic regionalization plan is already laid out along these two axes before 2 months gestation.

3.1 DORSO-VENTRAL PATTERNING OF THE NEURAL TUBE

Along the length of the developing neural tube, molecules already described—Shh and BMPs—induce cells to develop particular fates according to their dorso-ventral positions. During neurulation, secretion of Shh by the notochord and floorplate results in increased Shh concentrations in the ventral aspects of the neural tube (Cohen, 2003; Salie et al., 2005; Ten Donkelaar and Hori, 2006). At the rostral end of the tube, Shh influences formation of the ventral prosencephalon, the region that will become the brain. More caudally, Shh induces formation of motor neurons and other ventral cell types in the spinal cord. Shh is able to induce different cell types from the same neural tube precursor tissue because cells respond differently to Shh as a function of its concentration (Figure 12). Other molecular signals secreted by the notochord and floorplate on the ventral side of the tube include retinoic acid and the Wnt pathway factors noggin and chordin. On the dorsal side of the neural tube, BMPs and another TGF-β family member, dorsalin, are secreted by local tissues and by the roofplate, and continue to influence the development in cells of the dorsal spinal cord, including the neural crest (Kandel et al., 2000; Purves et al., 2001).

3.2 ROSTRO-CAUDAL PATTERNING OF THE NEURAL TUBE

As it closes, the neural tube bends along the rostro-caudal axis to form flexures that define major subdivisions. The rostral end, which will ultimately become the forebrain (prosencephalon) and the midbrain (mesencephalon) are separated from the developing hindbrain (rhombencephalon) by the

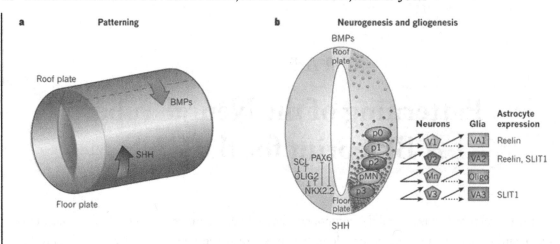

FIGURE 12: Dorso-ventral patterning of the neural tube (A) The primitive neuroepithelium of the neural tube is patterned by organizing signals. These signals emanate from the ventral floor plate (such as SHH, purple) and roof plate (BMPs and WNTs, green). (B) A cross-sectional view of the neural tube. Progenitors of motor neurons and interneurons are formed within distinct regionally restricted domains of the ventral neural tube: the p0, p1, p2, and p3 domains for interneuron subtypes, and the pMN domain for motor neurons. Dorsal domains are also similarly parcelled (not shown). Signaling mediated by SHH (gradient denoted by purple circles) regulates the expression of transcription factors (for example, NKX2.2, OLIG2, PAX6, and SCL) in the ventral neural tube. The interactions of these factors sharpen and maintain the domain boundaries. Reproduced with permission from Nature Publishing Group (Rowitch and Kriegstein 2010).

cephalic flexure, while junction of the hindbrain and spinal cord is delimited by the cervical flexure (Purves et al., 2001; Ten Donkelaar and Hori, 2006) (Figure 13). The prechordal plate at the rostral end of the embryo influences development of surrounding tissues to form the face and brain. Under the influence of the precordal plate, the prosencephalon balloons out from the rostral end of the neural tube (Figure 14a). Caudally, BMPs and Wnts are expressed at high levels; rostrally, molecules that bind and inactivate these signals enhance the gradient and prevent the prosencephalon from taking on a more caudal fate. The secreted signals turn on (and off) specific transcription factors, including OTX genes in the prosencephalon and HOX genes in the rhombencephalon (both types of homeobox containing transcription factor genes), whose expression levels depend on the concentration of signaling molecules to which a cell is exposed (Figure 14b,c). These transcription factors in turn are able to change the expression of other critical genes, including more transcription factors, leading to a cascade of changes that make different groups of cells along the length neural tube distinct. For example, in the hindbrain region, swellings called rhombomeres develop in which combinations of HOX genes define the identity of each segment (Kandel et al., 2000; Kiecker and

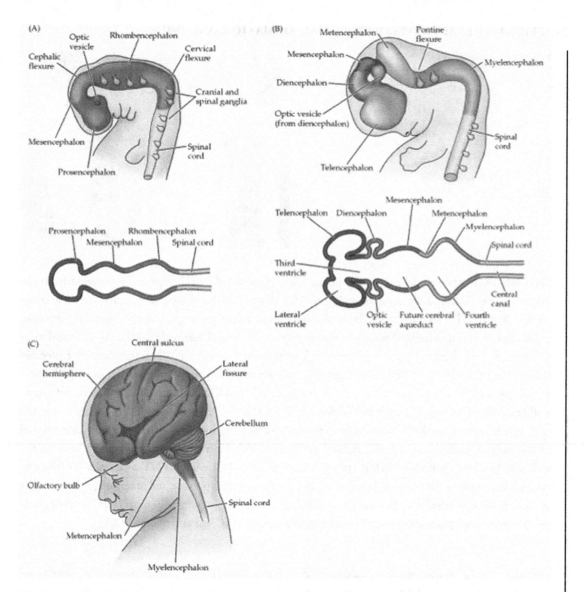

FIGURE 13: Regional specification of the developing brain. (A) Early in gestation the neural tube becomes subdivided into the prosencephalon (at the anterior end of the embryo), mesencephalon and rhombencephalon. The spinal cord differentiates from the more posterior region of the neural tube. The initial bending of the neural tube at its anterior end leads to a cane shape. Below is a longitudinal section of the neural tube at this stage, showing the position of the major brain regions. (B) Further development distinguishes the telencephalon and diencephalon from the prosencephalon; two other subdivisions—the metencephalon and myelencephalon—derive from the rhombencephalon. These subregions give rise to the rudiments of the major functional subdivisions of the brain, while the spaces they enclose eventually form the ventricles of the mature brain. Below is a longitudinal section of the embryo at the developmental stage shown in (B). (C) The fetal brain and spinal cord are clearly differentiated by the end of the second trimester. Several major subdivisions, including the cerebral cortex and cerebellum, are clearly seen from the lateral surfaces. Reproduced with permission from Sinauer Associates (Purves, Augustine et al. 2001).

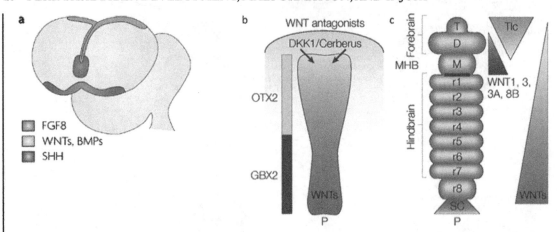

FIGURE 14: Rostro-caudal patterning of the telencephalon. (A) FGF8 (green) is secreted from the anterior neural ridge and the commissural plate; WNTs and BMPs (yellow) are secreted from the cortical hem; and SHH is secreted from the medial ganglionic eminence to generate positional information in the telencephalon. Modified with permission from (Dehay and Kennedy 2007). (B) A dorsal view of the entire neural tube depicts a posterior (P) to anterior gradient of WNT signaling that specifies posterior cell fates in the neural tube. Typically, anterior and posterior markers, such as orthodenticle homologue 2 (OTX2) and gastrulation brain homeobox 2 (GBX2), respectively, are used to determine A–P fates. WNT antagonists such as Dickkopf 1 (DKK1) and Cerberus restrict WNT signaling in the most anterior structures. DKK1 comes from the anterior axial mesoderm whereas Cerberus is expressed by the anterior endomesoderm. (C) A local gradient of WNTs creates signaling centers, such as the midbrain–hindbrain boundary (MHB), to regionalize the neural tube. The WNT antagonist Tlc (thought to be a homologue of the secreted Frizzled-related protein), which is expressed anteriorly, is required for the specification of telencephalic cells. D, diencephalon; r, rhombomere; SC, spinal cord; T, telencephalon. Modified with permission from Nature Publishing Group (Ciani and Salinas 2005).

Lumsden, 2005). In the prosencephalon, the compartments are less clear. However, the basic principle of graded signaling leading to patterns of transcription factors that specify regional identity are well conserved during brain development. In fact, the molecular divisions that form along this axis generate new boundary regions that themselves become signaling centers, adding complexity to the gradients of signals that further subdivide the prosencephalon.

3.3   PROSENCEPHALIC CLEAVAGE AND MIDLINE DEVELOPMENT

After the formation of the prosencephalon, it undergoes cleavage in three planes to produce paired structures. Horizontal cleavage produces the optic vesicles and olfactory bulbs and their corresponding tracts; transverse cleavage forms the telencephalon (which will ultimately develop into the cerebrum and basal ganglia) and the diencephalon (which will ultimately develop into a collection of separate forebrain structures including the thalamus and hypothalamus); and sagittal cleavage gives rise to the two cerebral hemispheres, the paired basal ganglia and lateral ventricles (Volpe, 2008). Prosencephalic midline development follows these cleavage steps to complete the overall structure of the human forebrain. These complex symmetric divisions again depend on gradients of signaling molecules, such as Shh, that trigger the differentiation of regional cell types via transcription factor induction. While complete failure of prosencephalic formation is rarely seen (because it is likely lethal early in gestation), failures in cleavage and midline development lead to relatively common human malformations.

3.4   MIDLINE DEFECTS: HOLOPROSENCEPHALY, AGENESIS OF THE CORPUS CALLOSUM, AND DANDY–WALKER MALFORMATION

3.4.1   Holoprosencephaly

Holoprosencephaly refers to complete or partial failure of normal prosencephalic cleavage processes (Geng and Oliver, 2009). In the most severe form, alobar holoprosencephaly, there is no division between the two cerebral hemispheres, with a single common ventricle, and fusion or absence of the midline structures (fused basal ganglia and thalamus, absence of the corpus callosum, olfactory bulbs and olfactory tracts, and absence or hypoplasia of the optic nerves) (Geng and Oliver, 2009; Volpe et al., 2009) (Figure 15). Because the face and brain are developing simultaneously under the influence of the notochord and prechordal mesoderm, holoprosencephaly may be accompanied by facial dysmorphisms; the most severe includes cyclopia (the presence of a single central eye) with the presence of a rudimentary nasal structure (proboscis); in such cases, holoprosencephaly is lethal. Facial manifestations in less severe forms include hypotelorism (abnormally closely set eyes), cleft palate and lip, and the presence of a single nostril, as shown in Figure 15 (Volpe, 2008; Geng and Oliver, 2009). Individuals with mildest forms may live a relatively normal life, with only mild cognitive impairments. Affected infants often have endocrinopathies, as the major endocrine organ of the brain, the hypothalamus, is a midline structure which also fails to form normally (Volpe et al., 2009).

Like the vast majority of brain malformations, holoprosencephaly has diverse etiologies, both environmental and genetic (Geng and Oliver, 2009). Monogenic forms of holoprosencephaly, which account for 15–20% of all cases (Volpe, 2008), are typically inherited in an autosomal dominant fashion. The majority of the mutations identified involve various players in the Shh pathway (Geng and Oliver, 2009), including Shh, patched homolog 1, transcription factor Gli-2, SIX3 (a recently discovered transcription factor required for Shh production) (Jeong et al., 2008) and Dispatched. In these monogenic, nonsyndromic familial forms, there is a remarkable phenotypic variability between individuals carrying the same mutation; parents of an infant or fetus with alobar holoprosencephaly may exhibit very subtle midline anomalies such as hypotelorism, hyposmia, absent frenulum,

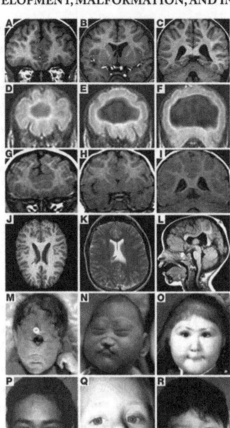

FIGURE 15: Clinical manifestations of holoprosencephaly. (A–I) Coronal images of control and HPE brains from anterior (A, D, and G) to posterior (C, F, and I). (A–C) In the control brain, the two hemispheres are separated completely (arrow in A) and the septum (arrow in B) and the corpus callosum (arrowhead in B) are present. (D–F) In alobar HPE, a single cerebral ventricle is present and the interhemispheric fissure is completely absent. (G–I) In semilobar HPE, the two hemispheres are incompletely separated (arrow in G) and the septum and corpus callosum are absent (arrow and arrowhead in H, respectively). (J and K) Horizontal images of control (J) and lobar HPE (K). The septum is present in the control brain (arrow in J); however, it is partially absent in the lobar HPE brain (arrow in K). (L) Sagittal image of a MIH brain. The genu and splenium of the corpus callosum are present (arrows in L); however, the corpus callosum is absent at the region lacking the interhemispheric fissure (arrowhead in L). (M–O) Craniofacial defects associated with HPE. (M) Alobar HPE with cyclopia and proboscis. (N) Semilobar HPE with microcephaly and cleft lip and palate. (O) Semilobar HPE with ocular hypotelorism and midface hypoplasia. (P and Q) Microforms of HPE. (P) Absence of nasal bones and cartilage with a narrow nasal bridge. (Q) Single central maxillary incisor. (R) MIH patient with normal facial appearance. Reproduced with permission from the American Society for Cinical Investigation (Geng and Oliver 2009).

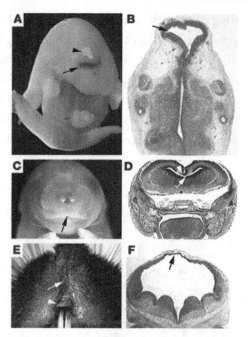

FIGURE 16: Mouse models of holoprosencephaly. (A and B) $Chd^{-/-}Nog^{-/-}$ embryo exhibiting alobar HPE–like phenotype: cyclopia (arrow in A) and proboscis (arrowhead in A). (B) Coronal section of $Chd^{-/-}Nog^{-/-}$ embryo highlighting the single cerebral ventricle (arrow). (C and D) $Six3^{+/ki}Shh^{+/-}$ embryos exhibit semilobar HPE–like phenotype: agenesis of philtrum (arrow in C), lack of corpus callosum (arrowhead in D), and a single telencephalic ventricle anteriorly (arrow in D). (D) Coronal section of a $Six3^{+/ki}Shh^{+/-}$ embryo. (E) Image of an adult $Cdo^{-/-}$ mouse exhibiting microforms of HPE: dysgenesis of philtrum (arrow) and single central maxillary incisor (arrowhead). (F) Coronal section of an $ShhN/+$ embryo exhibiting MIH-like phenotype: lack of dorsal telencephalic midline structures (arrow in F) and relatively normal ventral telencephalic structures. Reproduced with permission from the American Society for Cinical Investigation (Geng and Oliver 2009).

cognitive disability or simply a single maxillary central incisor (Hehr et al., 2004; Volpe, 2008). It is not yet understood what genetic and environmental modifiers of the Shh pathway account for this phenotypic variability. The development of mouse models with errors in the Shh pathway that recapitulate human holoprosencephaly (Figure 16) makes it possible to explore some of the modulating genetic or environmental factors that lead to this variation (Geng and Oliver, 2009).

3.4.2   Agenesis of the Corpus Callosum

Once the basic patterning of the prosencephalon is underway, several interior brain structures begin to form as tissue prominences emerging from its inner surface (Volpe et al., 2009). Here, we focus on the example of the corpus callosum, the largest interhemispheric white matter tract (commissure) in the brain ultimately, consisting of over 190 million axons (Paul et al., 2007) (Figure 17). Axons connecting homologous regions on each side of the brain are arranged topographically in the anteroposterior axis (Engle, 2010). When the corpus callosum forms, axons must cross from one side of the brain to the other. Multiple steps are required: a midline glial scaffold for axons forms; the first "pioneering axons" from the cingulate cortex cross and serve as a guide to axons that follow; and crossing axons must follow the pioneering axons via signaling from the interaction of cell adhesion molecules (CAMs) with the developing axonal growth cones (Paul et al., 2007).

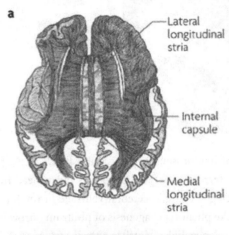

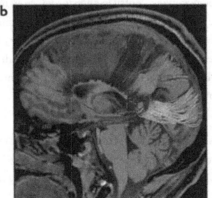

FIGURE 17: Organization of a human corpus callosum. (A) Anatomical drawing of corpus callosum viewed with cortex retracted. (B) Diffusion MRI imaging allows modeling of fiber tracts. Fibers are colored according to their projection areas: prefrontal lobe (green), premotor and supplementary motor areas (light blue), primary motor areas (dark blue), primary sensory cortex (red), parietal lobe (orange), occipital lobe (yellow), and temporal lobe (violet). Modified with permission from Nature Publishing Group (Paul, Brown et al. 2007).

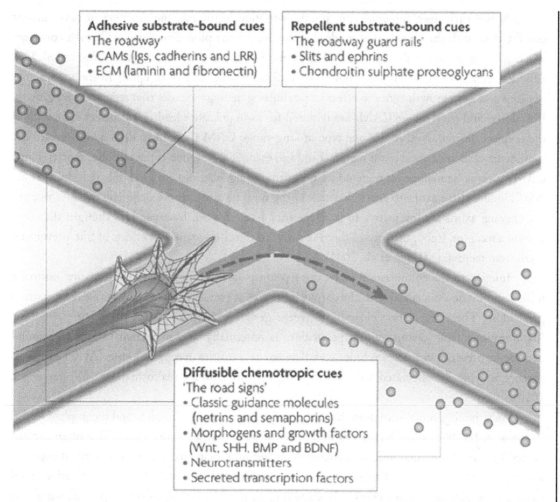

Adhesive substrate-bound cues
'The roadway'
• CAMs (Igs, cadherins and LRR)
• ECM (laminin and fibronectin)

Repellent substrate-bound cues
'The roadway guard rails'
• Slits and ephrins
• Chondroitin sulphate proteoglycans

Diffusible chemotropic cues
'The road signs'
• Classic guidance molecules
  (netrins and semaphorins)
• Morphogens and growth factors
  (Wnt, SHH, BMP and BDNF)
• Neurotransmitters
• Secreted transcription factors

FIGURE 18: Schematic of the growth cone environment that controls axon connections. The axon travels on a "road" that is made up of adhesive molecules that are either presented directly on a neighboring cell surface (for example, transmembrane cell adhesion molecules (CAMs)) or assembled into a dense and complex extracellular matrix (ECM; for example, laminin and fibronectin). Additionally, anti-adhesive surface-bound molecules (such as slits, ephrins, and chondroitin sulphate proteoglycans) can prohibit growth cone advance and thus provide the 'guard rails' that determine the road boundaries. Finally, diffusible chemotropic cues are the 'road signs' that present further steering instructions to the growth cone and include various diffusible chemotropic molecules (such as netrins and semaphorins), as well as morphogens (such as Wnt, sonic hedgehog (SHH) and bone morphogenetic protein (BMP)) and growth or neurotrophic factors (such as brain-derived neurotrophic factor (BDNF)) secreted transcription factors and neurotransmitters. Attraction (green) or repulsion (red) cues is not due to the particular signal, but rather to the specific growth cone receptors that are activated and the internal signaling of the growth cone. Modified with permission from Nature Publishing Group (Lowery and Van Vactor 2009).

At this early developmental time point, disruption in the midline cleavages or development can interfere with the signals that guide the crossing of the pioneering axons. Growth cones are specialized organelles at the tip of growing axons that respond to these signals (Figure 18). Cell adhesion molecules (CAMs) are extracellular molecules that bind to receptors in the membranes of axons and axon growth cones to effect intracellular signaling cascades that regulate axon growth. Both long- and short-range CAMs are required for axon guidance leading to correct formation of the corpus callosum. Netrins are one type of long-range CAM thought to draw axons toward the midline from their original brain region (Lindwall et al., 2007). Once at the midline, shorter-range CAMs such as Semaphorins, secreted by the surrounding "scaffold" of glial cells (Lindwall et al., 2007), induce these axons to cross glial cells at the midline also express a chemorepellant, Slit, but the crossing axons are insensitive to it. Once they have crossed, however, it is thought that they express a receptor, Robo, that makes them sensitive to the chemorepellant effects of Slit, preventing them from recrossing (Long et al., 2004).

Inferior to the commissural plate, the septum pellucidum, a smaller commissure, normally forms in intimate association with the corpus callosum, as two leaves referred to as a cavum septum pellucidum. These leaves fuse close to 40 weeks gestation (term) to form the mature structure. Identification of the cavum septum pellucidum is potentially very important in fetal ultrasound because its presence virtually excludes complete agenesis of the corpus callosum (ACC), while its absence is typically associated with a wide range of other brain malformations (Winter et al., 2010).

ACC is a brain malformation that frequently occurs in both complete and incomplete forms; it is seen in 1:1000–1:6000 births and in its mildest form may go undiagnosed. It is often accompanied by other brain anatomic abnormalities, including Chiari malformation described above in association with spina bifida. Prognosis varies; cases associated with other brain malformations tend to have poor outcomes, while cases of isolated total or partial agenesis of the corpus callosum may result in no or only mild neuropsychological impairment (Volpe, 2008; Volpe et al., 2009; Engle, 2010). Additional structures, called bundles of Probst, consisting of fibers that would normally form the corpus callosum but instead coarse posteriorly, may be seen (Figure 19), suggesting that the process of axon guidance failed prior to crossing (Volpe, 2008). Diffusion tensor imaging (DTI), a powerful method for tracing the course of axonal tracts in the brain, suggests that failures in axon guidance are often the cause of ACC, but the molecular underpinnings have not yet been elucidated (Paul et al., 2007; Engle, 2010).

3.4.3   Dandy–Walker Malformation

An additional midline structure that is prone to major malformations is the cerebellar vermis (Figure 20). The cerebellum is part of the hindbrain and plays a critical role in refining movements by

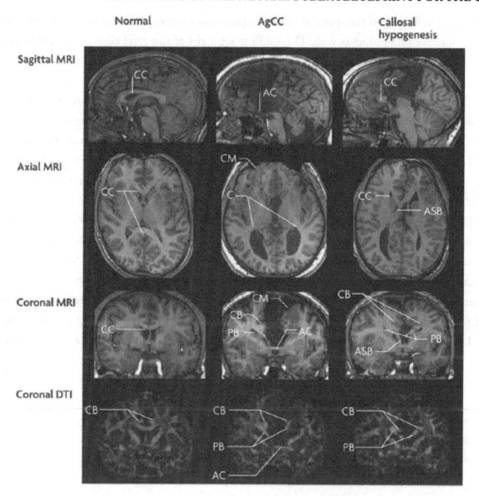

Nature Reviews | Neuroscience

FIGURE 19: Neuroanatomical features of agenesis of the corpus callosum and callosal hypogenesis revealed by MRI and diffusion tensor imaging (DTI). Structural T1-weighted MRI (top 3 rows) and directionally encoded color anisotropy dMRI (bottom row) are shown from a normal young adult male volunteer (left column), a young adult male with ACC (middle column), and a young adult male with callosal hypogenesis (right column). The DTI images encode fiber orientation in white matter tracts using a three-color scheme such that fiber pathways with predominantly left–right orientation are displayed as red, anteroposterior orientation as green, and craniocaudal orientation as purple. AC, anterior commissure; ASB, anterior sigmoid bundle; C, colpocephaly; CB, cingulum bundle; CC, corpus callosum; CM, cortical malformation; PB, Probst bundle. Reproduced with permission from Nature Publishing Group (Paul, Brown et al. 2007).

modulating signals sent to the spinal cord from the motor cortex; more recently, it has been implicated in cognitive processing as well. The midline segment of the cerebellum, the vermis, receives sensory input from various brain structures, allowing the cerebellum to adjust and refine precise motor movements.

The cerebellum develops from the dorsal aspect of the rhombencephalon from focal thickenings referred to as rhombic lips. Progenitor cells located in the rhombic lips generate neurons which migrate in radial planes to form the nuclei of the cerebellum, the Purkinje cells and the granule cells (Millen and Gleeson, 2008). Under the developing cerebellum, the choroid plexus forms in the fourth ventricle along with the midline foramen (foramen of Magendie) through which cerebrospinal fluid (CSF) flows (Volpe, 2008).

The Dandy–Walker syndrome (Figure 21) is a constellation of abnormalities that result from failure of vermian and foramen development to occur: complete or partial agenesis of the cerebellar vermis; dilation of the fourth ventricle; and enlargement of the posterior fossa with superior displacement of the tentorium (Volpe, 2008). The compression of midbrain structures by the cyst, together with the cerebellar vermis hypoplasia, leads to truncal ataxia, apneic spells, nystagmus and cranial nerve palsies; these symptoms are greatly improved with decompression of the cyst (Fenichel, 2009). Long-term neurodevelopmental outcomes vary greatly in children with Dandy–Walker mal-

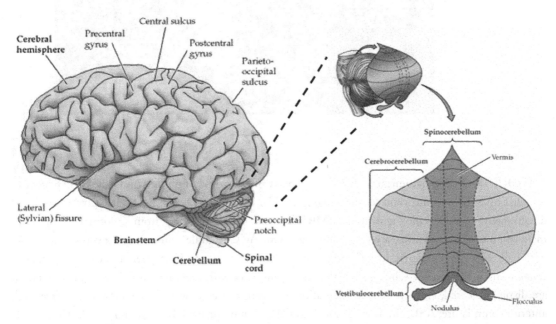

FIGURE 20: Anatomy of the cerebellum with flattened view of the cerebellar surface illustrating the three major subdivisions. Modified with permission from Sinauer Associates (Purves, Augustine et al. 2001).

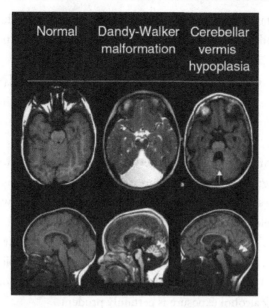

FIGURE 21: Cerebellar vermis malformations. Top row: Axial images at the level of the midbrain-hindbrain junction (isthmus). Bottom row: Midline sagittal images. In Dandy–Walker malformation (DWM), the posterior fossa is full of fluid (white field) and bottom shows the cystic dilatation of the fourth ventricle (*). Cerebellar vermis hypoplasia (CVH) shows reduced vermis without other associated features. Modified with permission from Elsevier (Millen and Gleeson 2008).

formation; the prognosis is best when there are no associated brain malformations and the vermis is relatively normal appearing, in which case some individuals may have a normal IQ (Bolduc and Limperopoulos, 2009).

3.5   GENETICS AND ENVIRONMENT COLLIDE AGAIN TO GENERATE MIDLINE ALTERATIONS

Perhaps not surprisingly, many of these midline defects occur in association with each other. In addition, they may share molecular underpinnings and/or be caused by associated environmental factors. For example, holoprosencephaly is linked to the hedgehog pathway and several genes linked to Dandy–Walker (Zic1 and Zic4) are regulated in the developing cerebellum by Shh. Current investigations are focusing on the role of primary cilia in both of these malformations. ACC can additionally be associated with NTDs, as in Meckle–Gruber syndrome, a rare syndrome with CNS abnormalities, renal cystic dysplasia and other anomalies. Environmental factors, such as exposure to alcohol prior to 3 months gestation, have also been linked to these patterning malformations. However, the effects of even chronic alcohol consumption are remarkably variable; maternal age

(>30 years), nutritional status and genetic background all appear to modulate the development of fetal alcohol syndrome and the more recently recognized brain malformations associated with it.

3.6   SUMMARY: CNS PATTERNING AND MIDLINE DEFECTS

- The layout of the CNS occurs by division of the neural tube into more specialized regions (along its dorso-ventral axis) and into two symmetric halves (along the rostro-caudal axis).
- The basic patterning of the central nervous system is established by the 3-dimensional interactions of signaling molecules expressed along these two major axis.
- Hedgehog pathway activity is one of the major molecular pathways required for midline division and development from the forebrain to the hindbrain.
- Major midline defects occur well before the end of the first trimester of pregnancy (3 months gestation) and the causes are again multifactorial (genetic, metabolic, environmental).
- Unlike NTDs, no treatments have been identified to reduce major midline malformations although limiting alcohol exposure is certainly prudent.

·   ·   ·   ·

CHAPTER 4

Neural Proliferation and Migration (3 Months Gestation Into Postnatal Period)

4.1   NEUROGENESIS

With the major structures of the brain in place, the next task is to populate it with cells. Neurogenesis, most prominent between the third and fourth months of gestation, is the generation of neuronal cells by mitosis from a population of relatively undifferentiated progenitor cells. This process begins first within the ventricular zone (VZ) and next in the subventricular zone (SVZ) (Figure 22). Initially, symmetric divisions of pluripotent neuroepithelial cells, the neural stem cells, divide at the ventricular surface to form additional pluripotent neural stem cells. Over the past decade, it has been demonstrated that these pluripotent cells are equivalent to radial glia, with a process spanning the widening telencephalon (Gotz and Barde, 2005). These neurogenic radial glia then undergo asymmetric, self-renewing divisions that generate additional progenitors. As the SVZ forms, it fills with intermediate progenitors that are rapid, transient amplifying cells. In this way, the radial glia give rise to proliferative units in which each neuronal cell is derived from one of the progenitor cells and thus shares a common lineage with the other neurons produced by the same progenitor. The final quantity of neurons in the brain is largely determined by the number of proliferative units and the number of divisions of the transient amplifying cells (Purves et al., 2001; Woods et al., 2005; Volpe, 2008).

In human cortical development, Kriegstein and colleagues (Hansen et al., 2010) have recently have recently identified an additional region of rapid proliferation in outer SVZ. In this outer SVZ, radial glia that have detached from the VZ divide and proliferate, then lose their pial contact and have an increased capacity for generating intermediate progenitors (Figure 23). A combination of neurogenic mechanisms is at work to rapidly generate billions of neurons during this brief period of development. While neurogenesis continues in specific brain regions even into adulthood, this period marks the neural expansion that characterizes the human cortex.

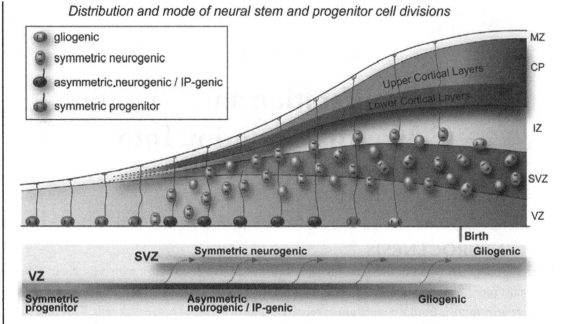

Distribution and mode of neural stem and progenitor cell divisions

FIGURE 22: Schema of neurogenesis in the dorsal telencephalon illustrating location, cleavage plane orientation, and mode of neural stem and progenitor cell divisions. Radial glial (RG) cells divide vertically at the surface of the ventricular zone (VZ) throughout cortical development. Before neurogenesis begins most RG divisions are symmetric self-renewing (red), expanding the founder population of VZ progenitor cells. At the onset of neurogenesis RG cells undergo asymmetric self-renewing divisions (dark blue) that produce either neurons or intermediate progenitor (IP) cells. Daughter neurons produced directly by RG cells may form the lower cortical layers. RG cells produce IP cells throughout the remainder of cortical neurogenesis. Daughter IP cells (light blue) divide close the ventricle during early stages of neurogenesis, but their location shifts away from the ventricle and the IP cells divide in the subventricular zone (SVZ) once that structure has formed. Most IP cells divide horizontally and produce symmetric pairs of neurons that form the upper cortical layers. After producing neurons RG cells translocate away from the ventricle and produce glial progeny (green). IZ, intermediate zone; MZ, marginal zone. Reproduced with permission from Wiley-Blackwell (Noctor, Martinez-Cerdeno et al. 2008).

While most neural proliferation occurs in the embryonic brain, the neurons of the adult brain are not all post-mitotic. In the past decades, adult neurogenesis has been recognized in rodents and primates. It occurs primarily in the dentate gyrus of the hippocampus and in the SVZ of the lateral ventricles, generating the rostral migratory stream of the olfactory bulbs. The extent to which adult neurogenesis plays a role in humans, either normally or after neurological damage, remains controversial and is under intense investigation (recently reviewed in Ming and Song, 2011).

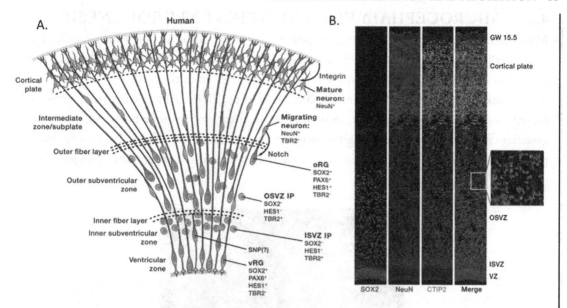

FIGURE 23: Human corticogenesis: expansion via a newly identified outer SVZ. A. Outer radial glia (oRG) cells, IP cells, and migrating neurons (red to green) present in the human outer subventricular zone (OSVZ) allow neural expansion. The number of ontogenetic "units" is significantly increased with the addition of oRG cells over ventricular radial glia (vRG) cells. Maintenance of oRG cells by Notch and integrin signaling is shown. Short neural precursors (SNP), a transitional cell form between RG and IP cells, are also depicted. B. Histology of human developing cortex demonstrating the outer OSVZ. Cells expressing the neuronal markers NeuN (RBFOX3, red) and CTIP2 (BCL11B, green) make up 45% of the OSVZ population (gestational week [GW] 15.5) but never co-label with the progenitor cell marker SOX2 (blue, inset). Modified with permission from Elsevier (Lui, Hansen et al. 2011).

4.2   PROGRAMMED CELL DEATH

The description of neural proliferation would not be complete without mentioning its flip side, programmed cell death (PCD). PCD, also called apoptosis, occurs throughout the developing cortex and eliminates up to 50% of neurons in any given area. Neuronal survival or death depends on the balance of signals the cell receives. The majority of PCD occurs in the late third trimester of gestation. During this period, cells are particularly vulnerable to having the PCD genetic mechanisms triggered. It is easy to imagine how an environmental insult during this period, such as an episode of hypoxia and/or ischemia, might trigger this cascade and injure particularly sensitive brain regions. When term infants are treated with hypothermia after an acute hypoxic–ischemic event (see below), it is this PCD cascade that may be interrupted leading to improved neurological outcomes.

4.3   MICROCEPHALY VERA: DEFICIENT NEUROGENESIS

Microcephaly vera, historically referred to isolated microcephaly, is an abnormally small, but architecturally normal brain, without other congenital malformation (Figure 24). Microcephaly is defined as an abnormally small brain (>2 standard deviations from the mean cranial size adjusted for age, sex and ethnicity) (Woods et al., 2005; Cox et al., 2006). Any disorder that slows brain growth causes microcephaly because the skull grows in direct proportion to brain expansion during gestation. Microcephaly vera is intriguing because the abnormally small brain is typically deficient particularly in gray matter, and often has a simplified gyral pattern, suggesting a decrease in the

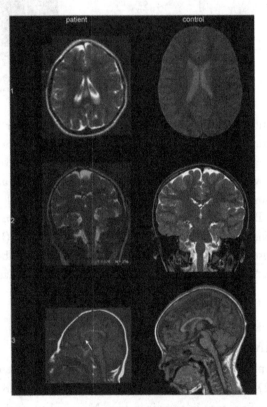

FIGURE 24: Microcephaly vera. Cranial MRI of the patient with an *ASPM* gene mutation causing microcephaly (A) compared to an age matched control (B). Axial (1) and coronal (2) T2-weighted images show the typical reduction of brain volume, especially of cerebral cortex in the patient in comparison to an age-matched control. Note the typical appearance of a simplified gyral pattern with scarce gyri in the patient. Sagittal images also illustrate the poorly developed frontal lobes in the patient and the agenesis of the rostrum of the corpus callosum (white arrow). Modified with permission from Elsevier (Kaindl, Passemard et al. 2010).

number of cerebral cortical neurons due to underproduction or premature loss (Kaindl et al., 2010). Some patients with microcephaly vera also have evidence of a migrational disorder.

Some genetic causes of autosomal recessive microcephaly vera (referred to as Microcephaly Primary Hereditary (MCPH)) have recently been identified. MCPH is rare, having been diagnosed in approximately 100 families worldwide (Kaindl et al., 2010) The first mutation identified was located in a gene encoding a protein which was subsequently called Microcephalin. Microcephalin is expressed in mouse embryos in the ventricular zone (the site of neurogenesis) and is induced by DNA damage, suggesting that microcephalin may play a role in cell cycle check points and DNA repair. Cell cycle check points occur to ensure that during periods of cellular division (such as neurogenesis), each step of the cell cycle of growth and division is completed properly. DNA repair mechanisms are closely linked with cell cycle check points because DNA damage incurred with cellular divisions must be repaired prior to subsequent divisions to ensure the integrity of the resulting daughter cells. Studies conducted in cultured cell lines, cells from human subjects with microcephalin mutations, and in fruit flies and mice carrying microcephalin mutations all suggest that microcephalin is necessary for proper DNA repair, chromosomal condensation, spindle formation and coordination of nuclear replication (Kaindl et al., 2010). Neuropathological studies also reveal evidence of insufficient neurogenesis in human MCPH subjects (Volpe, 2008). The other genes that cause MCPH also cause cell cycle defects: cyclin-dependent kinase 5 regulatory associated protein; abnormal spindle-like, microcephaly associated; centromeric protein J and SCL/TAL1-interrupting locus STIL. One fascinating possibility is that the genes that are disrupted in microcephaly might have been involved in the evolution of relatively large head size in humans versus their primate ancestors (and thus the associated advancements in cognitive abilities). Microcephaly genes are therefore also being studied as candidate genes in the evolution of the human brain (Kaindl et al., 2010).

4.4   NEURONAL MIGRATION

As neurogenesis proceeds, the new cells being produced must migrate to their specified brain regions. In the human brain this happens mainly between 3 and 6 months gestation. Radial glial cells serve as a scaffold for the radial migration, that is, migration perpendicular to the growing cortical plate (Figure 25). The neuronal cells that migrate radially will become the projection neurons of the brain—neurons with long axons that transmit information between different regions of the brain. Interneurons, by contrast, have short axons and transmit information within a single brain region (Martin, 1989). The interneuron precursors (primarily those that will be the GABAergic inhibitory interneurons) are primarily born in the ganglionic eminence, not the VZ or SVZ, and migrate tangentially into the cortex (McManus and Golden, 2005).

Radial migration, the dominant form of migration, was observed many decades prior to tangential migration. This radial migration lead to the "radial unit hypothesis" (Rakic, 1988) which

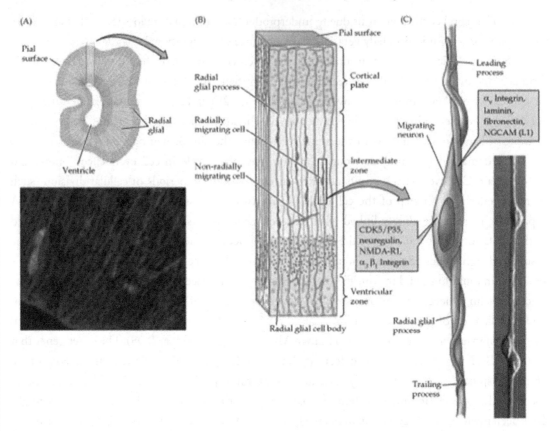

FIGURE 25: Radial migration in the developing cortex. (A) Section through the developing forebrain showing radial glial processes from the ventricular to the surfaces. Micrograph shows migrating neurons labeled with an antibody to neuregulin, specific for migrating cortical neurons. (B) Enlargement of boxed area in (A). Migrating neurons are intimately apposed to radial glial cells, which guide them to their final position in the cortex. Some cells take a nonradial migratory route, which can lead to wide dispersion of neurons derived from the same precursor. (C) A single neuroblast migrates upon a radial glial process (based on serial reconstruction of EM sections as well as in vitro assays of migration, as shown in the accompanying micrograph). Cell adhesion molecules found on the surface of either the neuron (green) or the radial glial process (tan) are indicated in the respective boxes. Reproduced with permission from Sinauer Associates (Purves, Augustine et al. 2001).

proposed that neurons which are produced within the same proliferative unit migrate together, ultimately giving rise to ontogenic columns of neurons in the cortex that carried a proto-map for cortical organization from the VZ into the expanding cortex (Figure 26). The radial unit hypothesis has since been modified to incorporate the newer observations of tangential migration as well as the occurrence of neurogenesis in the human outer SVZ, as described above (Lui et al., 2011) (Figure 23).

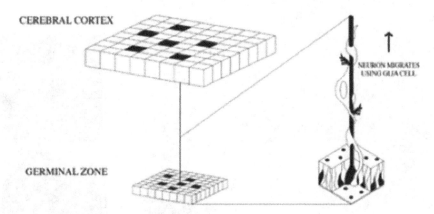

CEREBRAL CORTEX

GERMINAL ZONE

NEURON MIGRATES
USING GLIA CELL

FIGURE 26: Radial Unit Hypothesis (after Rakic, 1988). Proto-map for cortical organization was hypothesized to be carried from ventricular surface to cortex by radial glial ontogenic columns. This model has since been modified to include additional types of migration. See text for details. Modified with permission from Wiley-Blackwell (Spalice, Parisi et al. 2009).

The first migrating neurons come to rest in a growing layer within the preplate. The preplate was formed by a group of undifferentiated neuronal cells prior to the arrival of the majority of the migrating neurons, as early as 7 weeks gestation. This growing layer of migrating neurons forms the cortical plate which splits the preplate into the upper marginal zone (occupied by Cajal–Retzius neurons) and the lower subplate, both of which subsequently play critical roles in circuit organization. The preplate and Cajal–Retzius neurons play a critical role in providing molecular signals, such as a molecule called reelin, to migrating cells which indicate where they should stop (Meyer, 2010). The neuronal cells form the layers of the cerebral cortex in an "inside out" fashion, with the first cells to migrate comprising the deepest layers of the cortex (layer VI) and the last comprising layer II (McManus and Golden, 2005). The layers of the fetal forebrain have been identified as early as 4 months gestation in human pathology samples.

4.5   DISORDERED MIGRATION: LISSENCEPHALY AND LIS1

Lissencephaly, "smooth brain," is a disorder of neuronal migration with global neurologic consequences. It is characterized by failure of neurons to migrate radially to form the six-layered cerebral cortex; instead, a primitive and disordered four-layer cortex is present (Figure 27). On MRI, the disordered migration is evident as a lack of normal gyri on the surface of the brain (Figure 27). Affected infants present with generalized hypotonia and paucity of movement, develop progressive microcephaly and typically have mental retardation and severe seizure disorders; the disorder is often fatal (Fenichel, 2009).

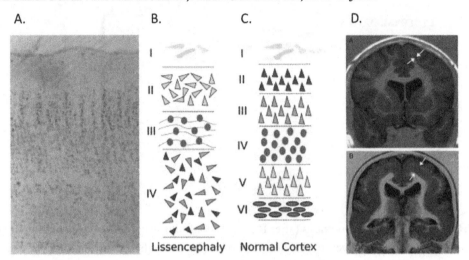

FIGURE 27: Lissencephaly (smooth brain). (A) Histopathological section from a patient with lissencephaly. Scheme of (B) cortical layering in classical lissencephaly (4 non-organized layers) and (C) normal cortical laminar organization (6 layers). (D) Brain MRI: coronal sections of the brain of two young women with different mutations of the *DCX* gene. Beneath the cortex, and separated from it by a thin layer of white matter, is present a subcortical laminar (band) heterotopia (white arrows) showing the same signal intensity as the cortex. The patient in B is more severely affected, as shown by the grade of simplification of the cortical pattern and the thickness of the heterotopic band. Modified with permission from Neurobiology of Disease (Guerrini and Parrini 2010).

Lissencephaly is most frequently caused by a loss of function mutation in the *LIS1* gene, that codes for a highly evolutionarily conserved protein (Wynshaw-Boris et al., 2010). LIS1 is part of an acetylhydrolase that degrades platelet-activating factor. LIS1 interacts with and stabilizes components of the cytoskeleton in their interaction with dynein cellular motors, and is necessary for coupling cytoskeletal and nuclear movement during cell migration (Wynshaw-Boris et al., 2010). When LIS1 function is lost, striking deficits in neuronal migration are observed (Tsai et al., 2005).

4.6   SUMMARY: NEURAL PROLIFERATION AND MIGRATION

- Neurogenesis begins at the ventricular surface with the division of neuroepithelial cells.
- Both symmetric and asymmetric divisions are used at different stages and in different locations (VZ versus SVZ) of the developing cortex.

- The large expansion in forebrain neurons seen in the human fetus depends on transient amplifying progenitors that vastly increase cell numbers.
- Post-mitotic neurons migrate into the cortex primarily along radial glia, although some neuronal types migrate tangentially.
- Human disorders of neural proliferation and migration lead to severe neurological deficits.

.   .   .   .

CHAPTER 5

Organization of Neuronal Circuits and Synaptogenesis (5 Months Gestation–Postnatal Years)

5.1 WIRING UP THE BRAIN: AXON PATH FINDING BETWEEN BRAIN REGIONS

The billions of neurons that are generated during development must become interconnected in precise neural circuits to function properly. How are these complex neural connections established? There are several steps that are required to shape the precise pattern of intricate connections that characterize the human brain. One key step is the guidance of axons to their target. Axonal guidance generally relies on genetically specified molecular cues that appear to provide for the reliable construction of a stereotyped framework. In Section 3.4, the need for axon pathfinding at the midline was briefly discussed in relation to ACC. As neurogenesis progresses, similar pathfinding needs to occur across many regions. To reach their correct destinations, axons are bundled together ("fasciculated") and directed to their targets by guidance cues in the extracellular environment (Raper and Mason, 2010). These cues can be attractive, pulling the axons toward their target, or repulsive, preventing axons from entering the wrong target area. The growth cone at the leading tip of the axon monitors and integrates pathfinding signals so that stereotyped connections between regions are established (Figure 28).

Four major families of guidance molecules have been identified—Netrins, Slits, Semaphorins and Ephrins—that act in concert with neural cell adhesion molecules (CAMs) and secreted signals (BMPs, Wnts and Shh) to get axons to their targets. Netrins are highly conserved molecules (found in worms, flies and mammals) that can act as both attractive and repulsive cues. Netrins were first isolated as factors secreted from the ventral midline floorplate that control commissural axon crossing at the midline (Serafini et al., 1996). Netrin receptors include Deleted in Colorectal Carcinomas (DCC) and UNC5, both are members of the immunoglobulin superfamily; the former primarily mediates attraction while the later mediates repulsion. At the midline of the spinal cord, axons meant to cross midline are initially attracted by Netrin, but once they cross midline they become unresponsive to Netrin; instead Slit protein expression repels them away from the midline, preventing recrossing. Slit actions are mediated receptors from the roundabout family (Robo), also immunoglobulin-related molecules. In addition to Netrins and Slits, two large families of guidance molecules, Semaphorins and Ephrins, can act as either attractive or repulsive cues for axon branching, as discussed in Section 3.4. Semaphorins and Ephrins are expressed not only at the midline but also in many linked regions. They have the astonishing flexibility to act either as signals or receptors depending on the context of their expression, adding even more flexibility to axon guidance systems.

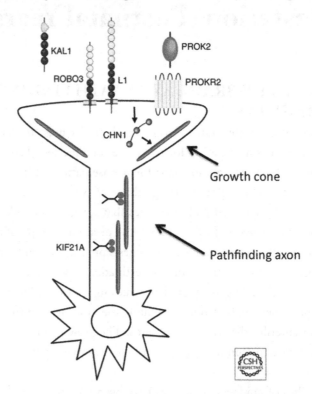

FIGURE 28: Schematic representation of gene products implicated in human disorders of axon guidance. ROBO3 (mutations responsible for HGPPS, see text), L1 (responsible for a form of hereditary spastic paraplegia), and PROKR2 (mutations responsible for an inherited form of anosmia) are shown as transmembrane receptors on the growth cone. KAL1 (anosmin) and PROK2 are shown as secreted ligands (mutations in these cause olfactory axon extension errors associated with Kallmann syndrome). CHN1 and KIF21 proteins both interact with microtubules and in these genes can cause types of ocular movement disorders due to ocular muscle innervation failures. Modified with permission from Cold Spring Harbor Laboratory Press (Engle 2010).

Few axon guidance molecules have been directly linked to human brain malformations, although some mutations have been implicated in clinical disease (Figure 28). Horizontal gaze palsy with progressive scoliosis (HGPPS) is very rare genetic disorder caused by a mutation in the *robo3* gene that encodes a Slit receptor. People with HGPPS are unable to move their eyes horizontally and develop spinal curvature. In HGPPS, the *robo3* mutation leads to a failure of motor and sensory pathways to cross the midline at the level of the brainstem. It remains unclear why HGPPS has such specific clinical findings when so many pathways are seemingly affected. While this and ACC discussed previously are the best-documented examples of axon pathfinding failures, additional pathfinding molecules, particularly in the Semaphorin pathway, have been linked to diverse neurodevelopmental disorders such as schizophrenia and epilepsy (reviewed in Yaron and Zheng, 2007).

5.2   THE ROLE OF SUBPLATE NEURONS IN DEFINING THE GROSS STRUCTURE OF CIRCUITS

While the future cortical neurons migrating into place and extending their axons, the subplate neurons, neurons that sit just below the neocortical plate, have matured and are helping to organize functional cortical circuits. Soon after their arrival below the cortical plate, subplate neurons differentiate and develop extensive and complex dendritic arbors. Axons from neurons whose cell bodies sit in the thalamus (the major relay station of the brain) extend into the subplate. The thalamic axons synapse onto the subplate neurons while waiting for their future targets, the cortical neurons, to migrate and mature. The axons of subplate neurons receiving thalamic afferent input then terminate on cells of the developing cortical plate and send collaterals back to thalamic neurons (Volpe, 2008). Subplate neurons express numerous receptors for neurotransmitters, neuropeptides and growth factors, reflecting their function in forming a "hub" which integrates signals from many sources and coordinates the formation of synchronized and functioning neural circuits (Kanold and Luhmann, 2010). Subplate neurons provide trophic support until the cortical plate is developed sufficiently to begin the process of forming connections directly with the thalamus. Once the neurons migrating to the cortical plate have formed the layers of the cortex, a period of rapid axonal growth occurs and marks the beginning of thalamocortical circuit formation. Here, the subplate neurons serve as guides, connecting thalamocortical, corticothalamic and corticocortical synapses correctly (Kanold et al., 2003; Kanold and Shatz, 2006; Kanold and Luhmann, 2010), Figure 29. Subplate neuron activity varies in different regions of the brain which develop in different stages; overall, the subplate neuron population peaks between 6 and 8 months gestation (Volpe, 2008). Once subplate neurons have served their function, most undergo apoptosis.

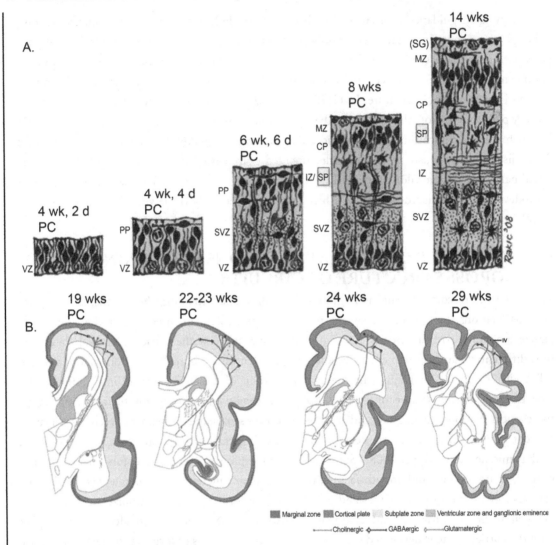

FIGURE 29: Subplate neurons are needed for thalamocortical wiring. A. 2008 revision of the Boulder Committee's schematic for development of the human cortex, highlighting the formation of the subplate by 8 weeks gestation. Preplate (PP); intermediate and subplate zones (IZ and SP); cortical plate (CP); intermediate zone (IZ); marginal zone (MZ); subventricular zone (SVZ); subpial granular layer (SG, part of the MZ); ventricular zone (VZ). B. Projections from thalamus to cortex by way of subplate in human fetuses and preterm infants. At 19 weeks postconception (PC) there is a 'waiting' period for the axons. In preterm infants at the edge of viability (22–23 PC), there is accumulation of afferents in the superficial subplate (i.e., below the cortical plate). Thalamic afferents penetrate into the cortical plate after 24 weeks PC to connect to layer 4. By 29 weeks PC, cortical elaboration of thalamocortical fibers. All major pathways 'cross' vulnerable periventricular zone. Note that afferent axons and cells in the subplate zone are involved in all stages of circuitry development and reorganization. Laminar shifts of thalamocortical

5.3 SYNAPTOGENESIS AND CRITICAL PERIODS: FINE CONTROL OF CIRCUITS

With the major structure of circuits in place at birth, these circuits are then further shaped and refined by the process of synapse formation. Synaptogenesis occurs at different times in different regions of the brain, but is generally characterized by a period of accelerated synapse formation until a peak number is achieved, after which point synaptic elimination occurs. Certain synapses will persist over time while others are terminated. In visual cortex, synaptogenesis peaks at 2–4 months after birth, maximum synapse numbers are achieved by the age of 8 months, and after elimination, 60% of the original number are present at 11 months. By contrast, in the frontal cortex, maximal synaptic density is reached at 15–24 months of age, and synapse elimination proceeds very gradually, occurring even into adolescence (Volpe, 2008).

When subcortical and cortical brain regions first connect, the initial pattern of connections is often imprecise. Precise connections are sculpted by elaboration, retraction and remodeling of axons and dendrites. These processes of elaboration, retraction and remodeling of neural connections within their targets is thought to a great extent to be "activity dependent" because the ultimate patterns of connections can be disrupted by blockade of neuronal activity (Katz and Shatz, 1996). The developmental periods of synaptic formation and elimination described above are referred to as "critical periods." Connections undergo substantial refinement during these developmental windows, and generally cannot be significantly altered by experience thereafter. Specific electrical activity in neurons is thought to fine-tune connections by adding or strengthening some neural connections and eliminating others.

What is the neural basis of these activity-driven changes in functional connections? This question has been actively studied in the visual system for more than 50 years (Figure 30). For example, spontaneous retinal activity *in utero* established the fundamental structure of thalamic layers (Penn et al., 1998) and cortical ocular dominance column (reviewed in Katz and Shatz, 1996). In the primary visual cortex, neurons are organized into columns which respond preferentially to input from one eye or the other. Classic experiments include suturing one eye closed (thereby eliminating all light input from that eye) fundamentally altered the ocular dominance columns. Neurons which, in the presence of binocular input would have preferentially responded to the sutured eye, instead began to respond strongly in response to the opposite (unsutured) eye (Figure 30B) (Hubel and Wiesel, 1964).

---

and basal forebrain afferents together with laminar development of synapses in the cortical plate cause changes in cortical electric responses. Note that weeks PC is 2 weeks less than the equivalent clinical gestational age (calculated from last menstrual period). A. modified with permission from Nature Publishing Group (Bystron, Blakemore et al. 2008). B. modified with permission from Collegium Antropologicum (Kostovic and Jovanov-Milosevic 2008).

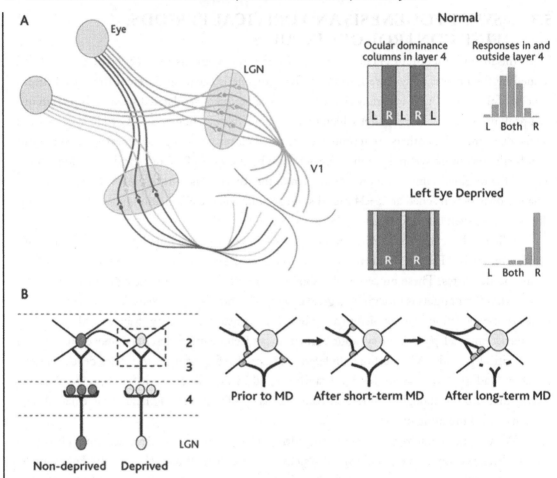

FIGURE 30: Activity refines visual cortex connection. Schematic summary of the refinement of visual cortical connections by neuronal excitation driven by vision: (A) Projections from the two retinas are targeted to the thalamic lateral geniculate nucleus (LGN) and subsequently to the primary visual cortex (V1). In higher mammals, the projections form alternating columns within layer 4, representing inputs from the right and left eyes, respectively. Suturing one eye shrinks its columns and causes cortical cells to respond nearly exclusively to the open eye. (B) Functional and structural changes after monocular deprivation (closure of one eye) occur rapidly in the superficial and deep layers of cortex. The first changes are a reduction in size and loss of spines (gray ovals) driven by the deprived eye. Deprivation for a longer period causes a more significant loss of spines driven by the deprived eye, shrinkage of deprived eye axon arbors (dashed line connecting upward), and an expansion of connections from the non-deprived eye. These changes occur during a critical period and then are difficult to reverse, potentially leading to long-term visual dysfunction. Modified with permission from the American Association for the Advancement of Science (AAAS) (Sur and Rubenstein 2005).

What are the molecular determinants of critical periods? Cholinergic neurotransmission play an important role in circuit modulation during critical periods, and even into adulthood. Afferents from basal forebrain neurons and cholinergic interneurons within the cortex provide cholinergic input to the developing cortex which stimulates waves of excitatory activity that are thought to induce neurite outgrowth and synaptogenesis (Bruel-Jungerman et al., 2011). Recently, a molecular "brake," Lynx1, was identified, which appears to slow synaptic plasticity at the end of the critical period in visual cortex by reducing cholinergic neurotransmission (Morishita et al., 2010). Mice lacking Lynx1 regained normal ocular dominance column structure even after monocular suture during the critical period, an effect that appears to result from enhanced cholinergic neurotransmission. Thus, an interesting possibility is that as further molecular determinants of critical periods are identified, they could be exploited for the purpose of regaining plasticity after injury in the developing brain, even outside of critical periods.

The synaptic plasticity present in the nervous system endows it with the ability to adapt to the many variations of the external world, such as in language acquisition during childhood or the ability to recover from a stroke when one region of cortex subsumes the functions of a damaged area. However, this plasticity also leaves the nervous system uniquely vulnerable to injury from abnormal patterns of activity, particularly during perinatal development.

5.4   IMMUNE MOLECULES AND CIRCUIT REFINEMENT

Immune molecules appear to play an important role in refining synapses as well (recently reviewed in Garay and McAllister, 2010). MHCI genes were first found to be expressed differentially during activity-dependent refinement of visual system connections (Corriveau et al., 1998). In addition, complement proteins have been implicated in synaptic refinement in the visual system and elsewhere in the developing CNS ((Boulanger et al., 2001); (Stevens et al., 2007)) (Figure 31). These immune molecules may help to translate neural activity, or lack thereof, into removal of inactive synapses. Interactions between these immune molecules, maternal infections and impaired synaptogenesis have been suggested to underlie neurodevelopmental disorders such as autism and schizophrenia, both of which are linked epidemiologically to maternal viral infections at the end of the first and beginning of the second trimester of gestation. How infection and circuit formation interact is a critical area of current investigation in perinatal brain injury.

5.5   IMPAIRED SYNAPSE FORMATION: FRAGILE X

Even with the gross structure of circuits in place, the importance of "fine tuning" these circuits through synaptogenesis and pruning is made evident by disorders in which this process is disrupted.

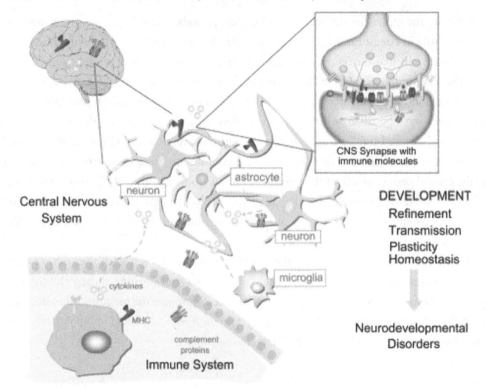

FIGURE 31: Complex relationship between the immune system and nervous system. However, it is now apparent that immune molecules not only cross the blood–brain barrier in times of injury, but are expressed during normal brain development. Recent evidence suggests roles for MHCI and its receptors, complement, and cytokines on the function, refinement, and plasticity of cortical and hippocampal synapses. These functions for immune molecules during neural development suggest that they could also mediate pathological responses to chronic elevations of cytokines in neurodevelopmental disorders, including autism spectrum disorders (ASD) and schizophrenia. Reproduced with permission from Frontiers Media (open access) (Garay and McAllister 2010).

Fragile X syndrome is the most common cause of mental retardation. Fragile X syndrome results from a trinucleotide repeat sequence in the FMR1 gene located on the X chromosome; the mutation precludes production of the FMR protein (FMRP). Males are more severely affected and exhibit a characteristic facial appearance, mental retardation and in some cases, autism (Fenichel, 2009). While gross brain malformations are not observed, histological inspection of spines from Fragile X patients reveal abnormally long and dense spines, an immature appearance that is consistent with a defect in experience dependent pruning or maturation of spines (Irwin et al., 2001) (Figure 32).

Dendritic spines are small protrusions from dendrites containing the post-synaptic elements of excitatory synapse and are pockets where chemical activity provoked by neurotransmission, including protein synthesis, can occur. They also provide one way to examine the developmental status of synapses histologically. Specifically, readily apparent spine morphology changes reflect changes in synapse and circuit properties, including those that occur with developmental synaptogenesis and pruning (Ethell and Pasquale, 2005). Spine morphology is known to be altered in many other forms

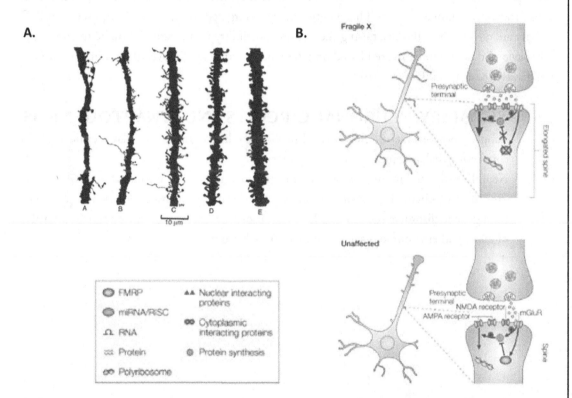

FIGURE 32: Dendritic spine abnormalities in Fragile X. A. Defects in FMRP result in long, thin immature dentritic spines. B. FMRP has a role in synaptic maturation and pruning, possibly through its regulation of gene products involved in synaptic development, and may have a regulatory role in activity-dependent translation at the synapse. Stimulation of postsynaptic metabotropic glutamate receptors (mGluRs) normally results in increased FMRP. mGlu activation also leads to internalization of AMPA receptors, inhibiting excitability, and FMRP dampens this effect. The absence of FMRP results in over-amplification of this response. A. Modified with permission from Sinauer Associates (Volpe 2008). B. Modified with permission from Nature Publishing Group (Gatchel and Zoghbi 2005).

of mental retardation including Rett syndrome (Armstrong, 2005), Down syndrome (Kaufmann and Moser, 2000) and Angleman syndrome (Jay et al., 1991).

These spine abnormalities have been replicated in mice lacking the FMR gene, which also display cognitive deficits and impaired synaptic plasticity. One theory which has garnered ample scientific support is that FMRP plays an important role in suppressing local protein synthesis in response to excitatory input from metabotropic glutamate receptors (mGluRs) (Meredith and Mansvelder, 2010; Dolen et al., 2007; Lee et al., 2008). Thus, FMRP is regulated by activity, suggesting that it may play a role in the refinement of neural circuits, a hypothesis that fits well with abnormal spine production. In the absence of FMRP, the protein synthesis goes unchecked when mGluRs are activated by neuronal activity. Therapeutic strategies targeted to suppressing the excess mGluR activity are now in clinical trials, raising the exciting possibility that Fragile X will be the first neurobehavioral disorder to be treated based on a therapeutic strategy discovered in animal models of the disease (Krueger and Bear, 2010).

5.6   SUMMARY: NEURONAL CIRCUITS AND SYNAPTOGENESIS

- Subplate neurons are the first cortical neurons to differentiate; they play a critical role in connecting thalamic projections to neocortex.
- Neural activity and immune molecules interact to sculpt synaptic connections.
- Disorders of circuit formation appear to underlie many neurodevelopmental disorders, both genetic disorders such as fragile X and disorders that seem to arise from an interplay of genes and environment such as autism and schizophrenia.

·   ·   ·   ·

CHAPTER 6

Gliogenesis and Myelination (5-Month Childhood)

6.1   GLIAL PROLIFERATION

Most of the cells that make up the human brain are not neurons, but rather are glia: astrocytes, oligodendrocytes and microglia. Glia provide trophic support for neurons, wrap axons in myelin to allow rapid conduction of action potentials and regulate synapse formation and plasticity. In mammals, gliogenesis follows neurogenesis. This switch from generating neurons to generating glia is critical because it determines the ultimate balance between neuron and glial numbers, and thus proper CNS function.

6.2   ASTROGENESIS

Astrocytes, the star-shaped glia, perform many functions: scaffolding, trophic support for neurons and for endothelium, maintenance of extracellular ion and water balance, and repair in response to injury. Astocytes are made by the SVZ progenitors, as well as arising directly from terminal differentiation of the radial glia (see Section 4.1 and Figure 22). The precise timing of astrocyte proliferation remains uncertain. Astrocytes are seen as early as 15 weeks gestation, but given that their production is associated with the terminal differentiation of radial glia, their genesis is most prominent in the human neocortex in the third trimester of gestation (Sen and Levison, 2006; Rowitch and Kriegstein, 2010). The factors that allow this switch from neural to astroglial fate are an active area of investigation. External signals, such as Wnts, seem to work in tandem with intrinsic genetic mechanisms, including methylation of glial-fate promoting genes, to control this switch (reviewed in Freeman, 2010). Immature astrocytes are vulnerable to hypoxia, ischemia and inflammation and may contribute substantially to perinatal brain injury (Sen and Levison, 2006).

6.3   OLIGOGLIOGENESIS AND MYELIN FORMATION

Oligodendrocytes produced in the third trimester of gestation and in the early postnatal period. These are the glia that produce myelin in the CNS. Compared to astrocytes and microglia, their development and role in perinatal brain injury is better understood (Rowitch and Kriegstein, 2010;

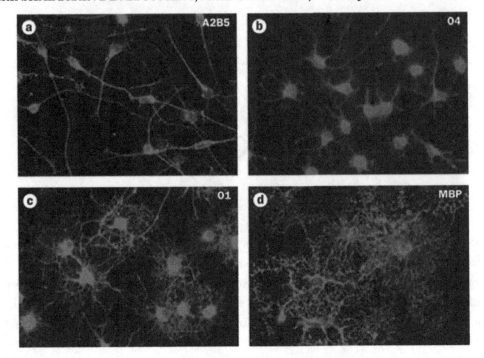

FIGURE 33: Stages of oligodendrocyte development. During oligodendrocyte maturation the cells undergo dramatic morphological changes and express sequential stage-specific markers. A. Early oligodendrocyte precursor cells express A2B5, B. whereas pre-oligodendrocytes express the antigen O4 and C. immature oligodendrocytes express the marker O1. D. Mature oligodendrocytes are distinguished from oligodendrocyte precursor cells and immature oligodendrocytes by their selective expression of myelin basic protein (MBP). Reproduced with permission from Nature Publishing Group (Deng 2010).

Silbereis et al., 2010). Four major stages of oligodendrocyte differentiation have been delineated (Figure 33). Oligodendrocyte progenitors (OPCs) originate from the SVZ starting at about 13 week gestation in humans. They are produced first ventrally and then progressively more dorsally as development continues. They proliferate and migrate, transitioning to pre-oligodendrocytes in the developing white matter regions starting at about 20 weeks gestation. Pre-oligodendrocytes are the predominant type in the periventricular white matter (the area around the lateral ventricles) during the second half of gestation. It is these cells that are most susceptible to injury following preterm birth (see Section 7.3).

The transition to myelinating cells is controlled by many of the signals already discussed, including Shh, BMPs, Wnts and other growth factors (Rosenberg et al., 2007). Myelination occurs over a prolonged period of time starting at about 7 months gestation in the thalamus and progressing into childhood. White matter matures progressively from central to peripheral and posterior to anterior. The association pathways of the frontal cortex are the last to myelinate during the transition from adolescence to adulthood.

6.4   MICROGLIA

Microglia are macrophage-like immune cells of the CNS that arise from bone marrow precursors. They are present in the developing white matter as early as the late first trimester of gestation. They remain primarily in the white matter during later gestation, suggesting that they may be contributors to neonatal white matter damage resulting from injury and/or infection. Microglia primarily act as scavenger cells, removing cellular debris. Activation of microglia occurs via Toll-like receptors (TLRs) that respond strongly to pathogen-associated molecules and trauma. Once activated, microglia trigger inflammation with associated release of cytokines that can damage oligodendrocytes and neuronal axons (reviewed in Chew et al., 2006).

6.5   SUMMARY: GLIOGENESIS

- Generation of glia is critical for normal functioning of the developing nervous system.
- Much of gliogenesis occurs after neurogenesis, but there is significant overlap during development.
- Many of the same signaling molecules that direct neurogenesis also direct gliogenesis.
- Regional and temporal differences in gliogenesis likely contribute specific vulnerabilities in the developing brain.

·   ·   ·   ·

CHAPTER 7

Developmental Brain Injury: Before, During, and After Birth

7.1   NEONATAL ENCEPHALOPATHY

Both preterm and term infants can experience neurological depression or "encephalopathy" (abnormal consciousness, seizures or abnormal tone and reflexes) after birth. In preterm infants (born at less than 37 weeks of gestation), this can extend over weeks, while in term infants, it is typically defined as occurring in the first three days of life. Both preterm and term neonatal encephalopathies (NE) indicate injury that can damage both neurons and glia; both preterm and term brain injuries contribute to the overall burden of cerebral palsy and mental retardation. However, the specific vulnerability of the brain at each point in gestation is determined by its developmental state—a complex and rapidly evolving state as reviewed in the sections above.

While there are many mechanisms of cell injury to which the CNS is vulnerable, it is worth briefly reviewing a few of the most common ones:

- Oxidative stress: the cumulative damage seen in response to cytotoxic oxidants and free radicals produced in cells. The developing brain has high concentrations of unsaturated fatty acids, high rates of oxygen consumption, low antioxidants concentrations and exposure to redox catalysts (particularly iron), making it particularly vulnerable to oxidative damage.
- Excitotoxicity: the process by which excessive excitatory neurotransmitter (glutamate and possibly GABA which can be depolarizing early in development) release allows high levels of calcium to enter cells via neurotransmitter receptors. Calcium activates enzymes that when mis-activated are destructive to cellular components. High expression of excitatory receptors and failure of immature glia to contain excitatory neurotransmitters put the developing brain at risk (Figure 34).
- Apoptosis: inappropriate triggering of the programmed cell death pathways when low levels of apoptotic inhibitors are present as seen in development. More severe insult can lead to necrosis. Susceptibility of the developing brain can vary by age, region, and gender.

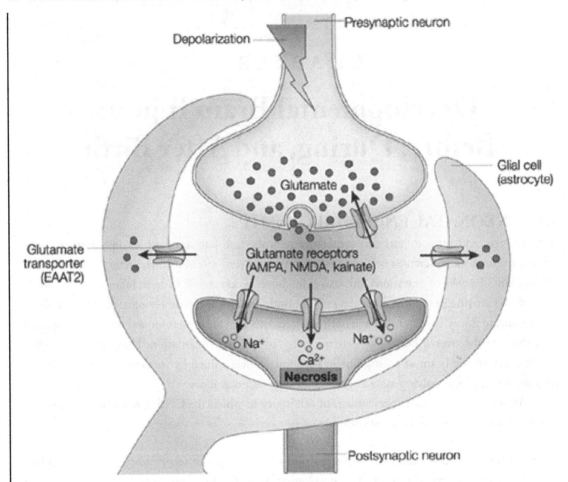

FIGURE 34: Excitotoxicity. Excitatory neurotransmitters such as glutamate are released from synapses on depolarization after the arrival of an action potential. The release process is carefully controlled and build-up of excessive neurotransmitter at the synapse is prevented by the action of dedicated transporters that clear the synaptic cleft. However, many deleterious conditions can converge to induce unrestrained glutamate release at synapses, initiating a cascade of events that leads to death of the postsynaptic cell. For example, catastrophic depolarization occurs during hypoxia or hypoglycaemia, which compromise energy production and therefore the ability of the cell to maintain a membrane potential. Reproduced with permission from Nature Publishing Group (Syntichaki and Tavernarakis 2003).

Inflammation accompanies each of these mechanisms, thus contributing to ongoing injury. Different combinations of these mechanisms can injure the developing brain, in utero, during birth or after delivery and the damage depends on insult and stage of development.

7.2   RISK FACTORS FOR INJURY BEFORE BIRTH: COMPROMISED PLACENTAL FUNCTION AND INFECTION

As the first functional fetal organ that exists at the interface between maternal and fetal circulation, the placenta monitors maternal well-being and nutritional status. Furthermore, it aims to modify this environment to optimize fetal well-being through changes in placental function. Substrate transport, metabolism and hormone production are all crucial placental functions, the failure of which may compromise fetal neurodevelopment. Impairment of any or all of these functions may contribute to NE either directly or by increasing fetal susceptibility to hypoxic–ischemic damage.

Prolonged failure of substrate transfer and hypoxia at the placental interface can directly cause growth failure that is strongly correlated with neurological compromise (Pardi et al., 2002; Jacobsson and Hagberg, 2004; Tan and Yeo, 2005; Jarvis et al., 2006; Jacobsson et al., 2008). Likewise, exposure to pro-inflammatory factors produced by the placenta itself in response to maternal infection have been linked to CP (Aaltonen et al., 2005; Keogh and Badawi, 2006; Maleki et al., 2009; Elovitz et al., 2011). Far less work has focused on connecting placental endocrine function to NE and CP because the fetus is often considered to develop in a "protected" environment, buffered from maternal and placental hormones. However, there are a growing number of reported associations between NE and abnormalities in hormones transported, modified or made by the placenta. For example, early thyroid hormone transferred from maternal circulation has long been recognized as required for normal neurogenesis, but several recent studies have also associated late maternal thyroid impairment with neonatal encephalopathy (Badawi et al., 2000; Girling and de Swiet, 2001; Kurinczuk et al., 2010; Locatelli et al., 2010). Exposure to glucocorticoids from maternal stress or iatrogenic exposure can change fetal hypothalamic–pituitary–adrenal axis response both pre- and postnatally, potentially contributing to CNS toxicity (recently reviewed in O'Donnell et al., 2009; Shinwell and Eventov-Friedman, 2009; Charil et al., 2010). Even small peptides, such as vasoactive intestinal peptide and oxytocin, have recently been shown in animal models to cross the placental barrier to modulate intrinsic neural excitability and neuronal survival (Zupan et al., 2000; Tyzio et al., 2006), implicating them in protection from perinatal excitotoxic damage.

Infection has been discussed in the sections above as contributing to inflammatory damage in the CNS and possibly interfering with normal functions of immune molecules that help shape neural connections (Section 5.4). Maternal infection may also act indirectly by changing placental function. Recent experiments in mice suggest that even mild innate immune activation in early pregnancy causes placental hemorrhages (Carpentier et al., 2011). Fetuses that survived this insult showed impaired fetal neurogenesis and other signs of CNS hypoxia. A combination of a maternal signaling pathway, acting via TLRs and inducing proinflammatory cytokine release, and a placental TNFα pathway (genetically belonging to the fetus) was critical for these impairments. Both direct and indirect actions of infection and inflammation influenced by both maternal and offspring's genetics may shape the ultimate function of that neonate's brain.

7.3 INJURY IN THE PRETERM BRAIN
7.3.1 Periventricular Leukomalacia

More than 50,000 infants weighing less than 1,500 g, typically below 32 weeks gestation, are born each year in the United States. More than 90% of these extremely low-birth weight infants will survive with modern neonatal intensive care. However, up to 10% will have cerebral palsy and up to 50% may have cognitive deficits. The lower the gestational age of the infant, the greater the risk of poor outcomes.

The "encephalopathy of prematurity" seen in these preterm infants is a complex combination of disrupted developmental events and direct injury (reviewed in Volpe, 2009) (Figure 35). These infants suffer from both frequent infection (often the immediate cause of the preterm delivery) and hypoxic–ischemic episodes due to poor lung function and immature hemodynamic control. The most common pathology seen in these preterm survivors is periventricular leukomalacia (PVL) (Figure 36), occurring in up to 50% of these infants. PVL can be focal—characterized by necrotic, cystic lesions in the white matter—or, as is seen more commonly in the past decade, can be diffuse—characterized by microscopic necrosis that results in small, diffuse glia scars in the white matter. Axon injury also occurs, possibly, as a sequelae of oligodendrocyte damage and microglial activation.

In the past decade, it has been recognized that PVL depends in large part on a maturation-dependent vulnerability of the pre-oligodendrocytes (Back et al., 2002). The pre-oligodendrocytes are the predominant oligodendrocyte type in the human brain between 20 and 32 weeks gestation—the key time of preterm brain vulnerability to PVL. These pre-oligodendrocytes are significantly more sensitive to excitotoxic, oxidative and inflammatory injury. Pre-oligodendrocytes do not have a well-developed antioxidant defense system and are therefore easily damaged by free radicals produced by ischemia as well as by activated microglia.

The decrease in oligodendrocytes in preterm brain injury is countered by a subsequent increase in pre-oligodendrocytes. However, these pre-oligodendrocytes seem to be a state of maturational arrest and do not mature into myelinating cells (reviewed in Volpe, 2009 and Silbereis et al., 2010). The cause of this maturation failure is under investigation, with the hope that it will be possible to mature these pre-oligodendrocytes so that they can remyelinate regions of diffuse PVL.

PVL is strongly associated with impairment of long-term brain growth and the neurode-

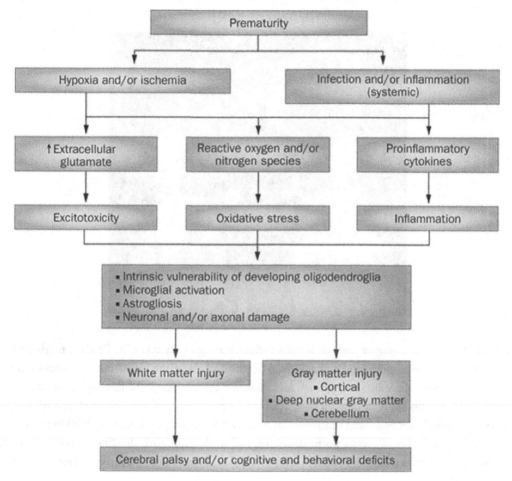

FIGURE 35: Risk factors for brain injury in the premature infant. Hypoxia–ischemia and infection and/or inflammation are the two main causes of brain injury in the developing brain. Excitotoxicity, oxidative stress and inflammation are the three main downstream mechanisms that cause injury to the developing brain. Neuropathological hallmarks include activated microglia, astrogliosis, and neuronal and/or axonal damage. Pre-oligo maturation arrest appears to be a critical step in PVL injury. The main clinical outcomes are cerebral palsy and cognitive and behavioral deficits. See text for details. Reproduced with permission from Nature Publishing Group (Deng 2010).

velopmental deficits seen in children who were born preterm (Ment et al., 2009). Abnormal myelination can be seen by advanced MRI imaging techniques or in *post mortem* histology. Focal PVL most commonly damages the motor tracts next to the lateral ventricles, resulting in spastic diplegia (impaired function of the legs more than the arms). Diffuse PVL is more commonly associated with long-term impairment of cognitive function.

Preterm MRI

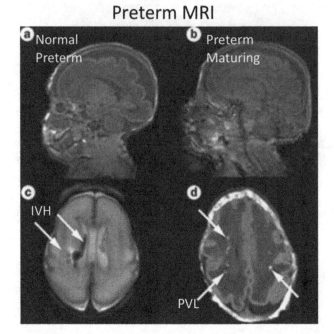

FIGURE 36: MRI imaging of preterm infants demonstrating injury types. A. Preterm newborn brain imaged at 26–27 weeks' postmenstrual age. B. MRI scan of the same infant taken 6 weeks after the scan in panel a. Note the substantial growth and increased complexity of the developing brain. C. MRI showing IVH in brain of preterm infant born at 29 weeks' postmenstrual age, 1 week after birth showing hemorrhage in the brain tissue and ventricles (arrows). D. Developing PVL seen on MRI brain scan of a preterm infant imaged at 30 weeks' postmenstrual age, 2 weeks after birth showing foci of white matter injury around ventricles (arrows). Modified with permission from Nature Publishing Group (Bonifacio, Glass et al. 2011).

7.3.2 Intraventricular Hemorrhage

Although its incidence has decreased in recent decades, one of the other major components of preterm brain injury is intraventricular and intraparenchymal hemorrhage (bleeding into the ventricles or into the brain tissue; IVH) (reviewed in McCrea and Ment, 2008) (Figure 36). Blood pressure fluctuations, respiratory distress, early gestational age at birth and infection all increase the risk of IVH, but the specific etiology remains unclear. The majority of IVH occurs within the first week of life. The germinal matrix, in which the precursor brain cells are generated, is very prominent in the preterm newborn particularly in the region of the caudate nucleus. Hemorrhage into this region with potential extension into the ventricles can disrupt ongoing gliogenesis, create inflammatory damage or lead to hydrocephalus (enlargement of the ventricles caused by impaired cerebrospinal fluid circulation). Intraparenchymal hemorrhage, although associated with severe IVH, is now recognized not to be extension of the blood in the ventricles into surrounding brain tissue, but rather bleeding into an area of venous infarct. IVH is frequently associated with the later appearance of PVL (McCrea and Ment, 2008; Volpe, 2008).

7.4   INJURY IN THE TERM BRAIN

7.4.1 Hypoxia–Ischemia

When NE is seen in a term newborn, it used to be automatically ascribed to hypoxia–ischemia, but it is now known that hypoxia–ischemia is only one of many possible causes. Long-standing pre-natal insults, such as the placental dysfunction discussed above, infection or metabolic disease can independently result in NE or contribute to a newborn's susceptibility to damage around the time of birth. Indeed, the majority of NE may be associated with problems that arise before the onset of labor. Since most of pregnancy is relatively unmonitored when compared to the hours around the delivery, most studies of term brain injury focus on hypoxic–ischemic events during this period; malpractice cases also focus on this period, despite the growing medical appreciation of antenatal factors in brain injury (ACOG, 2004).

NE can be called hypoxic–ischemic encephalopathy (HIE) however when it is due to an identifiable, acute hypoxic–ischemic episode. Acute intrapartum events, such as placental abruption, uterine rupture or a cord knot, can result in HIE, but are rare correlates of NE. More frequently, however, HIE is likely due to an acute injury superimposed on prior compromise (although para-doxically, prior insults can be protective by "pre-conditioning" the brain as well.) The damage from HIE is due not only to the loss of brain oxygenation followed by severe lactic acidosis (anaerobic metabolism) but also to secondary insults that include reperfusion injury, edema, failure of cerebral autoregulation and hemorrhage (Ferriero, 2004; Johnston et al., 2011).

Imaging studies of babies with HIE often shows damage to specific regions rather than global injury (Figure 37). Babies with severe encephalopathy may have deep basal ganglia injury and corti-cal injury, while babies with seizures alone often have focal hemorrhage. Severe hypoxia–ischemia that lasts less than 30 minutes, but that results in severe lactic acidosis, frequently leads to focal brain lesions in the peri-Rolandic cortex, ventrolateral thalamus and putamen. These children develop extrapyramidal CP that is characterized by impaired use of their hands compared with their legs, in contrast to the preterm infant's spatic diplegia (Ferriero, 2004; Johnston et al., 2011). The selective pattern of injury in term HIE suggests that some specific neuronal circuits are more vulnerable to injury at term, particularly those that are high in metabolic activity or rich in excitatory synapses.

HIE results in a cascade of the injurious events outlined: excitotoxicity, oxidative injury, cell death and ongoing inflammatory damage (Ferriero, 2004; Johnston et al., 2011). Many years of animal studies suggested that this cascade could be blocked by moderate hypothermia (brain cool-ing) to protect the brain tissue. In the past decade, clinical studies have shown that treating infants with moderate to severe HIE with hypothermia (cooling from normal body temperature of 37°C to 33°C) within 6 hours of birth for a duration of 72 hours improves survival without cerebral palsy or other disability by about 40% and reduces death or neurological disability by nearly 30%. Hypothermia for HIE has become standard in many neonatal intensive care units. Studies are in progress to optimize cooling protocols, as well as to identify adjuvant therapies that could help protect or repair the developing brain in the window afforded by cooling (Johnston et al., 2011).

7.4.2 Stroke

Perinatal stroke is an increasingly recognized cause of NE (Figure 37), occurring in about 1 in 4000 term births (Ferriero, 2004; Lynch, 2009). Most term infants with strokes present with seizures. The cause of the stroke is rarely identified. Maternal risk factors include prothrombotic (clot-

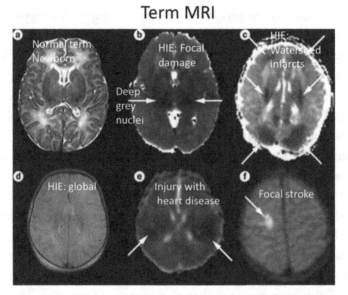

FIGURE 37: MRI imaging of term infants demonstrating injury types. A. Normal term brain MRI (T2-weighted image) B. Diffusion-weighted MRI scan showing hypoxic–ischemic injury to the deep gray nuclei (arrows). C. Diffusion-weighted MRI scan with watershed pattern of hypoxic–ischemic injury (top and bottom arrows point to anterior and posterior watershed regions of the cortex and white matter, central arrows show injury to white matter). D. T1-weighted imaged depicting diffuse brain injury secondary to global hypoxic–ischemic insult. E. Diffusion-weighted image showing injury to the white matter and cortex (arrows) in an 11-day-old term infant with congenital heart disease. F. Diffusion-weighted MRI scan showing focal stroke (arrow) in a term infant who presented with focal seizures on the second day of life. Modified with permission from Nature Publishing Group (Bonifacio, Glass et al. 2011).

generating) disorders, cocaine abuse, placental dysfunction or infection. Neonatal risk factors include prothrombotic disorders, congenital heart disease and infection. Arterial injury can occur during delivery, but are extremely rare. Perinatal stroke can lead to hemiparesis, cerebral palsy, epilepsy and cognitive problems, but development is normal in about 25% of these children, emphasizing the capacity for plasticity and repair in the developing brain.

7.5   SUMMARY: BRAIN INJURY

- Common mechanisms of cell injury exist across development.
- Manifestations of newborn brain injury are determined by the developmental stage of the brain as well as the type of insult.
- Although a mix of glia and neuronal damage occurs in injuries, preterm brain injury is primarily white matter damage while term brain injury is primarily specific regions of neuronal loss.

. . . .

CHAPTER 8

Conclusion

In the developing brain, the processes of neurulation, neural tube patterning and prosencephalic development, neurogenesis, neuronal migration, circuit formation and myelination begin in the first month of life and continue into adulthood. Genetic and environmental influences may intersect with each of these developmental stages to produce distinct forms of injury, as summarized in Table 1. Genetic causes include chromosomal abnormalities, single gene disorders, or mutations in multiple genes. Intrauterine environmental causes include infection, maternal ingestion of teratogenic drugs or other substances such as alcohol, maternal malnutrition and maternal metabolic abnormalities, such as diabetes. In some cases, a fetus with a genetic predisposition for a certain type

TABLE 1: Neurological disorders arising at developmental stages between embryogenesis and delivery.		
3–4 weeks gestation	Neurulation	Neural tube defects
2–3 months gestation	Prosencephalic development	Holoprosencephaly Agenesis of the corpus callosum
2–3 months gestation	Cerebellar development	Dandy–Walker malformation
3–4 months gestation	Neuronal proliferation	Microcephaly
3–5 months gestation	Neuronal migration	Lissencephaly
5–months gestation +	Circuit organization	Fragile X syndrome
Birth at 6 to 8 months gestation (preterm birth)	Myelination	Periventricular leukomalacia (white matter injury)
Neonatal encephalopathy at term	Regional susceptibility to injury	Cortical grey matter and deep grey nuclei damage

of brain injury may not develop the malformation except in the presence of certain environmental triggers, which may explain why there is often phenotypic heterogeneity between family members carrying identical gene mutations. In the case of prematurity, the "environmental factor" is simply being exposed to the external world while the brain continues development that normally occurs *in utero*.

The last few decades have provided great strides into the understanding of brain development and the pathogenesis of fetal brain injury; concurrently, the field of neonatology has made possible the survival of infants with various brain injuries. It is now up to the current generation of scientists to translate what we know about molecular mechanisms of fetal brain injury into meaningful therapies that will improve the prognosis of infants who begin life with a brain injury.

·   ·   ·   ·

References

Aaltonen, R., Heikkinen, T., et al. (2005). Transfer of proinflammatory cytokines across term placenta. *Obstet Gynecol* 106(4): pp. 802–7.

ACOG (2004). Neonatal encephalopathy and cerebral palsy: executive summary. *Obstetrics and Gynecology* 103(4): 780–1.

Adzick, N. S., Thom, E. A., et al. (2011). A randomized trial of prenatal versus postnatal repair of myelomeningocele. *N Engl J Med* 364(11): pp. 993–1004.

Armstrong, D. D. (2005). Neuropathology of Rett syndrome. *J Child Neurol* 20(9): pp. 747–53.

Back, S. A., Luo, N. L., et al. (2002). Arrested oligodendrocyte lineage progression during human cerebral white matter development: dissociation between the timing of progenitor differentiation and myelinogenesis. *J Neuropath Exp Neurol* 61(2): pp. 197–211.

Badawi, N., Kurinczuk, J. J., et al. (2000). Maternal thyroid disease: a risk factor for newborn encephalopathy in term infants. *BJOG* 107(6): pp. 798–801.

Blom, H. J., Shaw, G. M., et al. (2006). Neural tube defects and folate: case far from closed. *Nat Rev Neurosci* 7(9): pp. 724–31.

Blumenfeld, H. (2002). Neuroanatomy Through Clinical Cases. Sunderland, Sinauer Associates, Inc.

Bolduc, M. E. and Limperopoulos, C. (2009). Neurodevelopmental outcomes in children with cerebellar malformations: a systematic review. *Dev Med Child Neurol* 51(4): pp. 256–67.

Boulanger, L. M., Huh, G. S., et al. (2001). Neuronal plasticity and cellular immunity: shared molecular mechanisms. *Curr Opin Neurobiol* 11(5): pp. 568–78.

Bruel-Jungerman, E., Lucassen, P. J., et al. (2011). Cholinergic influences on cortical development and adult neurogenesis. *Behav Brain Res*.

Carpentier, P. A., Dingman, A. L., et al. (2011). Placental TNF-alpha signaling in illness-induced complications of pregnancy. *Am J Pathol* 178(6): pp. 2802–10.

Charil, A., Laplante, D. P., et al. (2010). Prenatal stress and brain development. *Brain Res Rev* 65(1): pp. 56–79.

Chew, L. J., Takanohashi, A., et al. (2006). Microglia and inflammation: impact on developmental brain injuries. *Ment Retard Dev Disabil Res Rev* 12(2): pp. 105–12.

Cohen, M. M., Jr. (2003). The hedgehog signaling network. *Am J Med Genet A* 123A(1): pp. 5–28.

Copp, A. J., Greene, N. D., et al. (2003). The genetic basis of mammalian neurulation. *Nat Rev Genet* 4(10): pp. 784–93.

Corriveau, R. A., Huh, G. S., et al. (1998). Regulation of class I MHC gene expression in the developing and mature CNS by neural activity. *Neuron* 21(3): pp. 505–20.

Cox, J., Jackson, A. P., et al. (2006). What primary microcephaly can tell us about brain growth. *Trends Mol Med* 12(8): pp. 358–66.

De Robertis, E. M. and Kuroda, H. (2004). Dorsal-ventral patterning and neural induction in Xenopus embryos. *Annu Rev Cell Dev Biol* 20: pp. 285–308.

Dolen, G., Osterweil, E., et al. (2007). Correction of fragile X syndrome in mice. *Neuron* 56(6): pp. 955–62.

Elovitz, M. A., Brown, A. G., et al. (2011). Intrauterine inflammation, insufficient to induce parturition, still evokes fetal and neonatal brain injury. *Int J Dev Neurosci*.

Engle, E. C. (2010). Human genetic disorders of axon guidance. *Cold Spring Harb Perspect Biol* 2(3): p. a001784.

Ethell, I. M. and Pasquale, E. B. (2005). Molecular mechanisms of dendritic spine development and remodeling. *Prog Neurobiol* 75(3): pp. 161–205.

Fenichel, G., Ed. (2007). Neonatal Neurology. Philadelphia, Churchill Livingstone Elsevier.

Fenichel, G. (2009). Clinical Pediatric Neurology: A Signs and Symptoms Approach. Philadelphia, Saunders Elsevier.

Ferriero, D. M. (2004). Neonatal brain injury. *N Engl J Med* 351(19): pp. 1985–95.

Freeman, M. R. (2010). Specification and morphogenesis of astrocytes. *Science* 330(6005): pp. 774–8.

Freese, J. L., Pino, D., et al. (2010). Wnt signaling in development and disease. *Neurobiol Dis* 38(2): pp. 148–53.

Garay, P. A. and McAllister, A. K. (2010). Novel roles for immune molecules in neural development: implications for neurodevelopmental disorders. *Front Synaptic Neurosci* 2: p. 136.

Geng, X. and Oliver, G. (2009). Pathogenesis of holoprosencephaly. *J Clin Invest* 119(6): pp. 1403–13.

Girling, J. and de Swiet, M. (2001). Maternal thyroid disease: a risk factor for newborn encephalopathy. *BJOG* 108(7): pp. 769–70.

Gotz, M. and Barde, Y. A. (2005). Radial glial cells defined and major intermediates between embryonic stem cells and CNS neurons. *Neuron* 46(3): pp. 369–72.

Greene, N. D., Stanier, P., et al. (2009). Genetics of human neural tube defects. *Hum Mol Genet* 18(R2): pp. R113–29.

Hansen, D. V., Lui, J. H., et al. (2010). Neurogenic radial glia in the outer subventricular zone of human neocortex. *Nature* 464(7288): pp. 554–61.

Hehr, U., Gross, C., et al. (2004). Wide phenotypic variability in families with holoprosencephaly and a sonic hedgehog mutation. *Eur J Pediatr* 163(7): pp. 347–52.

Huangfu, D. and Anderson, K. V. (2005). Cilia and Hedgehog responsiveness in the mouse. *Proc Natl Acad Sci U S A* 102(32): pp. 11325–30.

Huangfu, D., Liu, A., et al. (2003). Hedgehog signalling in the mouse requires intraflagellar transport proteins. *Nature* 426(6962): pp. 83–7.

Hubel, D. H. and Wiesel, T. N. (1964). Effects of Monocular Deprivation in Kittens. *Naunyn Schmiedebergs Arch Exp Pathol Pharmakol* 248: pp. 492–7.

Irwin, S. A., Patel, B., et al. (2001). Abnormal dendritic spine characteristics in the temporal and visual cortices of patients with fragile-X syndrome: a quantitative examination. *Am J Med Genet* 98(2): pp. 161–7.

Jacobsson, B., Ahlin, K., et al. (2008). Cerebral palsy and restricted growth status at birth: population-based case-control study. *BJOG* 115(10): pp. 1250–5.

Jacobsson, B. and Hagberg, G. (2004). Antenatal risk factors for cerebral palsy. *Best Pract Res Clin Obstet Gynaecol* 18(3): pp. 425–36.

Jarvis, S., Glinianaia, S. V., et al. (2006). Cerebral palsy and intrauterine growth. *Clin Perinatol* 33(2): pp. 285–300.

Jay, V., Becker, L. E., et al. (1991). Puppet-like syndrome of Angelman: a pathologic and neurochemical study. *Neurology* 41(3): pp. 416–22.

Jeong, Y., Leskow, F. C., et al. (2008). Regulation of a remote Shh forebrain enhancer by the Six3 homeoprotein. *Nat Genet* 40(11): pp. 1348–53.

Johnston, M. V., Fatemi, A., et al. (2011). Treatment advances in neonatal neuroprotection and neurointensive care. *Lancet Neurology* 10(4): pp. 372–82.

Kaindl, A. M., Passemard, S., et al. (2010). Many roads lead to primary autosomal recessive microcephaly. *Prog Neurobiol* 90(3): pp. 363–83.

Kandel, E. R., S. J. H., Jessell, T. M. (2000). Principles of Neural Science, McGraw-Hill.

Kanold, P. O., Kara, P., et al. (2003). Role of subplate neurons in functional maturation of visual cortical columns. *Science* 301(5632): pp. 521–5.

Kanold, P. O. and Luhmann, H. J. (2010). The subplate and early cortical circuits. *Annu Rev Neurosci* 33: pp. 23–48.

Kanold, P. O. and Shatz, C. J. (2006). Subplate neurons regulate maturation of cortical inhibition and outcome of ocular dominance plasticity. *Neuron* 51(5): pp. 627–38.

Katz, L. C. and Shatz, C. J. (1996). Synaptic activity and the construction of cortical circuits. *Science* 274(5290): pp. 1133–8.

Kaufmann, W. E. and Moser, H. W. (2000). Dendritic anomalies in disorders associated with mental retardation. *Cereb Cortex* 10(10): pp. 981–91.

Keogh, J. M. and Badawi, N. (2006). The origins of cerebral palsy. *Curr Opin Neurol* 19(2): pp. 129–34.

Kiecker, C. and Lumsden, A. (2005). Compartments and their boundaries in vertebrate brain development. *Nat Rev Neurosci* 6(7): pp. 553–64.

Krueger, D. D. and Bear, M. F. (2010). Toward fulfilling the promise of molecular medicine in fragile X syndrome. *Annu Rev Med* 62: pp. 411–29.

Kurinczuk, J. J., White-Koning, M., et al. (2010). Epidemiology of neonatal encephalopathy and hypoxic-ischaemic encephalopathy. *Early Hum Dev* 86(6): pp. 329–38.

Lee, J. L., Billi, F., et al. (2008). Wear of an experimental metal-on-metal artificial disc for the lumbar spine. *Spine (Phila Pa 1976)* 33(6): pp. 597–606.

Lindwall, C., Fothergill, T., et al. (2007). Commissure formation in the mammalian forebrain. *Curr Opin Neurobiol* 17(1): pp. 3–14.

Locatelli, A., Incerti, M., et al. (2010). Antepartum and intrapartum risk factors for neonatal encephalopathy at term. *Am J Perinatol* 27(8): pp. 649–54.

Logan, C. V., Abdel-Hamed, Z., et al. (2011) Molecular genetics and pathogenic mechanisms for the severe ciliopathies: insights into neurodevelopment and pathogenesis of neural tube defects. *Mol Neurobiol* 43(1): pp. 12–26.

Long, H., Sabatier, C., et al. (2004). Conserved roles for Slit and Robo proteins in midline commissural axon guidance. *Neuron* 42(2): pp. 213–23.

Lui, J. H., Hansen, D. V., et al. (2011). Development and evolution of the human neocortex. *Cell* 146(1): pp. 18–36.

Lynch, J. K. (2009). Epidemiology and classification of perinatal stroke. *Semin Fetal Neonatal Med* 14(5): pp. 245–9.

Maleki, Z., Bailis, A. J., et al. (2009). Periventricular leukomalacia and placental histopathologic abnormalities. *Obstet Gynecol* 114(5): pp. 1115–20.

Martin, J. H. (1989). Neuroanatomy: Text and Atlas. New York, Elsevier Science Publishing Company, Inc.

McCrea, H. J. and Ment, L. R. (2008). The diagnosis, management, and postnatal prevention of intraventricular hemorrhage in the preterm neonate. *Clin Perinatol* 35(4): pp. 777–92, vii.

McManus, M. F. and Golden, J. A. (2005). Neuronal migration in developmental disorders. *J Child Neurol* 20(4): pp. 280–6.

Ment, L. R., Hirtz, D., et al. (2009). Imaging biomarkers of outcome in the developing preterm brain. *Lancet Neurol* 8(11): pp. 1042–55.

Meredith, R. M. and Mansvelder, H. D. (2010). STDP and mental retardation: dysregulation of dendritic excitability in Fragile X syndrome. *Front Synaptic Neurosci* 2: 10.

Meyer, G. (2010). Building a human cortex: the evolutionary differentiation of Cajal–Retzius cells and the cortical hem. *J Anat* 217(4): pp. 334–43.

Millen, K. J. and Gleeson, J. G. (2008). Cerebellar development and disease. *Curr Opin Neurobiol* 18(1): pp. 12–9.

Ming, G. L. and Song, H. (2011). Adult neurogenesis in the mammalian brain: significant answers and significant questions. *Neuron* 70(4): pp. 687–702.

Morishita, H., Miwa, J. M., et al. (2010). Lynx1, a cholinergic brake, limits plasticity in adult visual cortex. *Science* 330(6008): pp. 1238–40.

Niehrs, C. (2004). Regionally specific induction by the Spemann-Mangold organizer. *Nat Rev Genet* 5(6): pp. 425–34.

O'Donnell, K., O'Connor, T. G., et al. (2009). Prenatal stress and neurodevelopment of the child: focus on the HPA axis and role of the placenta. *Dev Neurosci* 31(4): pp. 285–92.

Pardi, G., Marconi, A. M., et al. (2002). Placental-fetal interrelationship in IUGR fetuses—a review. *Placenta* 23 Suppl A: pp. S136–41.

Paul, L. K., Brown, W. S., et al. (2007). Agenesis of the corpus callosum: genetic, developmental and functional aspects of connectivity. *Nat Rev Neurosci* 8(4): pp. 287–99.

Penn, A. A., Riquelme, P. A., et al. (1998). Competition in retinogeniculate patterning driven by spontaneous activity. *Science* 279(5359): pp. 2108–12.

Purves, D. A., G. J., Fitzpatrick, D., Katz, L. C., LaMantia, A-S, McNamara, J. O., Williams, S. M., Ed. (2001). Neuroscience. Sunderland, Sinauer Associates, Inc.

Rakic, P. (1988). Specification of cerebral cortical areas. *Science* 241(4862): pp. 170–6.

Raper, J. and Mason, C. (2010). Cellular strategies of axonal pathfinding. *Cold Spring Harb Perspect Biol* 2(9): p. a001933.

Rosenberg, S. S., Powell, B. L., et al. (2007). Receiving mixed signals: uncoupling oligodendrocyte differentiation and myelination. *Cell Mol Life Sci: CMLS* 64(23): pp. 3059–68.

Rowitch, D. H. and Kriegstein, A. R. (2010). Developmental genetics of vertebrate glial-cell specification. *Nature* 468(7321): pp. 214–22.

Salie, R., Niederkofler, V., et al. (2005). Patterning molecules; multitasking in the nervous system. *Neuron* 45(2): pp. 189–92.

Sen, E. and Levison, S. W. (2006). Astrocytes and developmental white matter disorders. *Ment Retard Dev Disabil Res Rev* 12(2): pp. 97–104.

Serafini, T., Colamarino, S. A., et al. (1996). Netrin-1 is required for commissural axon guidance in the developing vertebrate nervous system. *Cell* 87(6): pp. 1001–14.

Shinwell, E. S. and Eventov-Friedman, S. (2009). Impact of perinatal corticosteroids on neuromotor development and outcome: review of the literature and new meta-analysis. *Semin Fetal Neonatal Med* 14(3): pp. 164–70.

Silbereis, J. C., Huang, E. J., et al. (2010). Towards improved animal models of neonatal white matter injury associated with cerebral palsy. *Dis Model Mech* 3(11–12): pp. 678–88.

Spemann, H. and Mangold, H. (2001). Induction of embryonic primordia by implantation of organizers from a different species. 1923. *Int J Dev Biol* 45(1): pp. 13–38.

Stevens, B., Allen, N. J., et al. (2007). The classical complement cascade mediates CNS synapse elimination. *Cell* 131(6): pp. 1164–78.

Tan, T. Y. and Yeo, G. S. (2005). Intrauterine growth restriction. *Curr Opin Obstet Gynecol* 17(2): pp. 135–42.

Ten Donkelaar, H. L. L., M., Hori A., Ed. (2006). Clinical Neuroembryology. Heidelberg, Germany, Springer-Verlag.

Traiffort, E., Angot, E., et al. (2010) Sonic Hedgehog signaling in the mammalian brain. *J Neurochem* 113(3): pp. 576–90.

Tsai, J. W., Chen, Y., et al. (2005). LIS1 RNA interference blocks neural stem cell division, morphogenesis, and motility at multiple stages. *J Cell Biol* 170(6): pp. 935–45.

Tyzio, R., Cossart, R., et al. (2006). Maternal oxytocin triggers a transient inhibitory switch in GABA signaling in the fetal brain during delivery. *Science* 314(5806): pp. 1788–92.

Volpe, J. J. (2008). Neurology of the Newborn, 5th Edition, Saunders Elsevier.

Volpe, J. J. (2009). The encephalopathy of prematurity—brain injury and impaired brain development inextricably intertwined. *Semin Pediatr Neurol* 16(4): pp. 167–78.

Volpe, P., Campobasso, G., et al. (2009). Disorders of prosencephalic development. *Prenat Diagn* 29(4): pp. 340–54.

Winter, T. C., Kennedy, A. M., et al. (2010). The cavum septi pellucidi: why is it important? *J Ultrasound Med* 29(3): pp. 427–44.

Woods, C. G., Bond, J., et al. (2005). Autosomal recessive primary microcephaly (MCPH): a review of clinical, molecular, and evolutionary findings. *Am J Hum Genet* 76(5): pp. 717–28.

Wynshaw-Boris, A., Pramparo, T., et al. (2010). Lissencephaly: mechanistic insights from animal models and potential therapeutic strategies. *Semin Cell Dev Biol* 21(8): pp. 823–30.

Yaron, A. and Zheng, B. (2007). Navigating their way to the clinic: emerging roles for axon guidance molecules in neurological disorders and injury. *Dev Neurobiol* 67(9): pp. 1216–31.

Zupan, V., Nehlig, A., et al. (2000). Prenatal blockade of vasoactive intestinal peptide alters cell death and synaptic equipment in the murine neocortex. *Pediatr Res* 47(1): pp. 53–63.

Author Biographies

Juliet Knowles is currently a pediatric resident and future pediatric neurology trainee at Lucile Packard Children's Hospital at Stanford. She graduated from Stanford in 2011 with an MD and PhD in Neuroscience. Her doctoral research, performed in the laboratory of Frank Longo, focused on the role of the p75 neurotrophin receptor in Alzheimer's disease. She was awarded a 2009 Alzheimer's Association Award for Young Scientists for her work.

Anna Penn is a neonatologist and developmental neurobiologist at Stanford University School of Medicine. She received her MD and PhD from Stanford University, was a pediatric resident at UCSF, and then returned to Stanford for fellowship training. She is an NIH-funded investigator, holding a 5-year NIH Director's New Innovator Award, as well as a recipient of multiple foundation grants. In her translational research, she draws from the many areas in which she has been trained—neuroscience, developmental biology, physiology, signal transduction, and neonatology—to develop novel models that reveal the impact of hormones on fetal brain development and damage. Her laboratory is investigating key steroid and peptide hormones that shape the developing brain in fetal and neonatal life, including an array of placental hormones and hormones that generate sex differences in the brain's response to damage.